Test-Driven Development with Java

Create higher-quality software by writing tests first with SOLID and hexagonal architecture

Alan Mellor

BIRMINGHAM—MUMBAI

Test-Driven Development with Java

Group Product Manager: Gebin George

Publishing Product Manager: Arvind Sharma

Senior Editor: Nisha Cleetus

Technical Editor: Jubit Pincy

Copy Editor: Safis Editing

Project Coordinator: Manisha Singh

Proofreader: Safis Editing

Indexer: Subalakshmi Govindhan

Production Designer: Shankar Kalbhor

Business Development Executive: Kriti Sharma

Marketing Coordinator: Sonakshi Bubbar

First published: January 2023

Production reference: 1231222

Published by Packt Publishing Ltd.
Livery Place
35 Livery Street
Birmingham
B3 2PB, UK.

ISBN 978-1-80323-623-0

www.packt.com

In memory of my mum, Eva Mellor (1928 – 2022). You saw me start this book but not finish it. If I'm perfectly honest, you wouldn't have enjoyed it as much as your Georgette Heyer novels.

– Alan Mellor

Contributors

About the author

Alan Mellor is an academy lead at BJSS, training the next generation of consulting software engineers, and the author of *Java OOP Done Right: Create object oriented code you can be proud of with modern Java*. Alan started with a Sinclair ZX81 computer with 1K of RAM and is very happy to have better computers now. Alan's work includes industrial control in C, web applications for e-commerce, gaming and banking in Java and Go, and document warehousing in C++. His most visible code is part of Nokia Bounce and the RAF Red Arrows flight simulator if you go back far enough.

I want to thank my wife Stephanie without whose support this book would not have been possible. I'm grateful to everyone who has taught me about software engineering, whether in person or via their books. All my love to Jake and Katy. You two are awesome.

About the reviewers

Jeff Langr has been building software professionally for over 4 decades. He's recognized as the author of five books on software development, including *Modern C++ Programming with Test–Driven Development: Code Better, Sleep Better, Agile in a Flash* (with Tim Ottinger), and *Agile in a Flash: Speed-learning Agile Software Development*. Jeff is also a co-author of the best-selling book *Clean Code*. He's written over a hundred published articles and several hundred blog posts on his site (https://langrsoft.com).

Nikolai Avteniev started his professional career at JPMorgan Chase, participated in the Extreme Programming Pilot, and learned how to apply test-driven development and continuous integration. After graduating from NYU with a degree in computer science, he took the experience of building and running an Agile development team to Real Time Risk Systems as one of the founding engineers. Nikolai later joined New York City AdTech start-up Intent Media and then moved on to building software teams and systems at LinkedIn (https://engineering.linkedin.com/blog/2017/08/getting-to-know-nikolai-avteniev).

Additionally, Nikolai teaches software engineering at the City University of New York. Currently, he works at Stripe, helping grow the GDP of the internet safely.

Table of Contents

Preface   **xv**

Part 1: How We Got to TDD

1

Building the Case for TDD   **3**

Writing code badly	3	Coupling and cohesion	10
Understanding why bad code is written	5	Decreasing team performance	12
Recognizing bad code	6	Diminishing business outcomes	14
Bad variable names	6	Summary	16
Bad function, method, and class names	6	Questions and answers	16
Error-prone constructs	9	Further reading	16

2

Using TDD to Create Good Code   **17**

Designing good quality code	17	Preventing logic flaws	25
Say what you mean, mean what you say	18	Protecting against future defects	27
Take care of the details in private	19	Documenting our code	28
Avoid accidental complexity	20	Summary	29
Revealing design flaws	22	Questions and answers	29
Analyzing the benefits of writing tests before production code	23	Further reading	30

3

Dispelling Common Myths about TDD — 31

Writing tests slows me down   31
Understanding the benefits of slowing down   32
Overcoming objections to tests
slowing us down   33

Tests cannot prevent every bug   34
Understanding why people say tests cannot
catch every bug   34
Overcoming objections to not catching
every bug   35

How do you know the tests are right?   35
Understanding the concerns behind writing
broken tests   35
Providing reassurance that we test our tests   36

TDD guarantees good code   36
Understanding problem-inflated expectations   36

Managing your expectations of TDD   37

Our code is too complex to test   37
Understanding the causes of untestable code   37
Reframing the relationship between good
design and simple tests   38
Managing legacy code without tests   39

I don't know what to test until I write
the code   39
Understanding the difficulty of starting with
testing   40
Overcoming the need to write production
code first   40

Summary   41
Questions and answers   41
Further reading   41

Part 2: TDD Techniques

4

Building an Application Using TDD — 45

Technical requirements   45
Preparing our development environment   46
Installing the IntelliJ IDE   46
Setting up the Java project and libraries   46

Introducing the Wordz application   48
Describing the rules of Wordz   49

Exploring agile methods   50

Reading user stories – the building block of
planning   51
Combining agile development with TDD   52

Summary   53
Questions and answers   53
Further reading   54

5

Writing Our First Test — 55

Technical requirements	55	Preserving encapsulation	64
Starting TDD: Arrange-Act-Assert	56	Learning from our tests	65
Defining the test structure	56	A messy Arrange step	65
Working backward from outcomes	58	A messy Act step	66
Increasing workflow efficiency	59	A messy Assert step	66
Defining a good test	59	Limitations of unit tests	66
Applying the FIRST principles	60	Code coverage – an often-meaningless metric	67
Using one assert per test	61	Writing the wrong tests	67
Deciding on the scope of a unit test	61	Beginning Wordz	67
Catching common errors	62	Summary	73
Asserting exceptions	63	Questions and answers	73
Only testing public methods	64		

6

Following the Rhythms of TDD — 75

Technical requirements	75	Writing our next tests for Wordz	79
Following the RGR cycle	75	Summary	92
Starting on red	76	Questions and answers	92
Keep it simple – moving to green	77	Further reading	93
Refactoring to clean code	78		

7

Driving Design – TDD and SOLID — 95

Technical requirements	96	Simplified future maintenance	100
Test guide – we drive the design	96	Counter-example – shapes code that violates SRP	101
SRP – simple building blocks	97	Applying SRP to simplify future maintenance	102
Too many responsibilities make code harder to work with	99	Organizing tests to have a single responsibility	104
Ability to reuse code	100	DIP – hiding irrelevant details	104

Applying DI to the shapes code   106

LSP – swappable objects   109

Reviewing LSP usage in the shapes code   111

OCP – extensible design   112

Adding a new type of shape   114

ISP – effective interfaces   115

Reviewing ISP usage in the shapes code   115

Summary   117

Questions and answers   117

8

Test Doubles – Stubs and Mocks   119

Technical requirements   119

The problems collaborators present for testing   120

The challenges of testing unrepeatable behavior   120

The challenges of testing error handling   121

Understanding why these collaborations are challenging   121

The purpose of test doubles   122

Making the production version of the code   124

Using stubs for pre-canned results   126

When to use stub objects   127

Using mocks to verify interactions   127

Understanding when test doubles are appropriate   130

Avoiding the overuse of mock objects   130

Don't mock code you don't own   130

Don't mock value objects   131

You can't mock without dependency injection   131

Don't test the mock   132

When to use mock objects   132

Working with Mockito – a popular mocking library   133

Getting started with Mockito   133

Writing a stub with Mockito   133

Writing a mock with Mockito   141

Blurring the distinction between stubs and mocks   142

Argument matchers – customizing behavior of test doubles   142

Driving error handling code with tests   144

Testing an error condition in Wordz   145

Summary   147

Questions and answers   148

Further reading   148

9

Hexagonal Architecture –Decoupling External Systems   149

Technical requirements   150

Why external systems are difficult   150

Environmental problems bring trouble   151

Accidentally triggering real transactions from tests   151

What data should we expect?   152

Operating system calls and system time   152

Challenges with third-party services   152

Dependency inversion to the rescue   153

Generalizing this approach to the hexagonal
architecture   154
Overview of the hexagonal architecture's
components   155
The golden rule – the domain never connects
directly to adapters   159
Why the hexagon shape?   160

Abstracting out the external system   160

Deciding what our domain model needs   160

Writing the domain code   164

Deciding what should be in our
domain model   164
Using libraries and frameworks in the
domain model   165
Deciding on a programming approach   165

Substituting test doubles for
external systems   166

Replacing the adapters with test doubles   166

Unit testing bigger units   167

Unit testing entire user stories   167

Wordz – abstracting the database   168

Designing the repository interface   168
Designing the database and random numbers
adapters   173

Summary   173

Questions and answers   174

Further reading   174

10

FIRST Tests and the Test Pyramid   175

Technical requirements   175

The test pyramid   176

Unit tests – FIRST tests   178

Integration tests   179

What should an integration test cover?   180
Testing database adapters   181
Testing web services   182
Consumer-driven contract testing   183

End-to-end and user acceptance tests   184

Acceptance testing tools   186

CI/CD pipelines and test
environments   187

What is a CI/CD pipeline?   187

Why do we need continuous integration?   187
Why do we need continuous delivery?   189
Continuous delivery or continuous
deployment?   190
Practical CI/CD pipelines   190
Test environments   191
Testing in production   192

Wordz – integration test for our
database   194

Fetching a word from the database   194

Summary   197

Questions and answers   197

Further reading   198

11

Exploring TDD with Quality Assurance — 199

TDD – its place in the bigger quality picture — 199
Understanding the limits of TDD — 200
No more need for manual testing? — 200

Manual exploratory – discovering the unexpected — 201

Code review and ensemble programming — 203

User interface and user experience testing — 205

Testing the user interface — 205
Evaluating the user experience — 207

Security testing and operations monitoring — 208

Incorporating manual elements into CI/CD workflows — 209

Summary — 210

Questions and answers — 211

Further reading — 211

12

Test First, Test Later, Test Never — 213

Adding tests first — 213
Test-first is a design tool — 214
Tests form executable specifications — 215
Test-first provides meaningful code coverage metrics — 215
Beware of making a code coverage metric a target — 216
Beware of writing all tests upfront — 217
Writing tests first helps with continuous delivery — 218

We can always test it later, right? — 218
Test-later is easier for a beginner to TDD — 219
Test-later makes it harder to test every code path — 219
Test-later makes it harder to influence the software design — 220
Test-later may never happen — 221

Tests? They're for people who can't write code! — 221
What happens if we do not test during development? — 222

Testing from the inside out — 222

Testing from the outside in — 224

Defining test boundaries with hexagonal architecture — 226
Inside-out works well with the domain model — 226
Outside-in works well with adapters — 227
User stories can be tested across the domain model — 228

Summary — 230

Questions and answers — 230

Further reading — 231

Part 3: Real-World TDD

13

Driving the Domain Layer   235

Technical requirements   235

Starting a new game   235
Test-driving starting a new game   236
Tracking the progress of the game   238
Triangulating word selection   244

Playing the game   249
Designing the scoring interface   250
Triangulating game progress tracking   252

Ending the game   254

Responding to a correct guess   255
Triangulating the game over due to too many
incorrect guesses   256
Triangulating response to guess
after game over   257
Reviewing our design   260

Summary   263
Questions and answers   264
Further reading   264

14

Driving the Database Layer   267

Technical requirements   267
Installing the Postgres database   267

Creating a database integration test 268
Creating a database test with DBRider   269
Driving out the production code   272

**Implementing the WordRepository
adapter**   276
Accessing the database   277
Implementing GameRepository   279

Summary   280
Questions and answers   280
Further reading   281

15

Driving the Web Layer   283

Technical requirements   283

Starting a new game   284
Adding required libraries to the project 284
Writing the failing test   285

Creating our HTTP server   288
Adding routes to the HTTP server   289
Connecting to the domain layer   290
Refactoring the start game code   293

Handling errors when starting a game   294

Fixing the unexpectedly failing tests   296

Playing the game   298

Integrating the application   304

Using the application   306

Summary   309

Questions and answers   310

Further reading   310

Index   313

Other Books You May Enjoy   322

Preface

Modern software is driven by users wanting new features, with no defects, released rapidly – a very challenging task. Solo developers have given way to development teams working together on a single software product. Features are added in short iterative cycles, then released to production frequently – sometimes daily.

Achieving this requires excellence in development. We must ensure that our software is always ready to be deployed and free of defects when it is released into production. It must be easy for our developer colleagues to work with. Code must be easy for anyone to understand and change. When the team makes those changes, we must have confidence that our new features work properly and that we have not broken any existing features.

This book brings together proven techniques that help make this a reality.

Test-driven development (TDD), the SOLID principles, and hexagonal architecture enable developers to engineer code that is known to work and known to be easy to work with. Development focuses on the fundamentals of software engineering. These practices are the technical foundation behind a code base that is easy and safe to change and always ready to be deployed.

This book will enable you to write well-engineered, well-tested code. You will have confidence that your code works as you think it should. You will have the safety of an ever-growing suite of fast-running tests, keeping a watchful eye over the whole code base as the team makes changes. You will learn how to organize your code to avoid difficulties caused by external systems such as payment services or database engines. You will reduce your dependence on slower forms of testing.

You will be writing higher quality code, suitable for a Continuous Delivery approach.

Modern software requires a modern development approach. By the end of this book, you will have mastered the techniques to apply one.

Who this book is for

This book is primarily aimed at developers who are familiar with the Java language basics, and who want to be effective in a high-performance Agile development team. The techniques described in this book enable your code to be delivered to production with few defects, and a structure that can be easily and safely changed. This is the technical basis of agility.

The book's early chapters will also be useful to business leaders who want to understand the costs and benefits of these approaches before committing to them.

What this book covers

Chapter 1, Building the Case for TDD, provides an understanding of the benefits TDD brings and how we got here.

Chapter 2, Using TDD to Create Good Code, covers some general good practices that will help us create well-engineered code as we apply TDD.

Chapter 3, Dispelling Common Myths about TDD, is a review of the common objections to using TDD that we might encounter, with suggestions for overcoming them. This chapter is suitable for business leaders who may have reservations about introducing a new technique into the development process.

Chapter 4, Building an Application Using TDD, involves setting up our development environment to build the Wordz application using TDD. It reviews how to work in short iterations using user stories.

Chapter 5, Writing Our First Test, introduces the basics of TDD with the Arrange, Act and Assert template. Writing the first test and production code for Wordz, we will look in detail at how TDD promotes a design step before we write code. We will consider various options and trade-offs, and then capture these decisions in a test.

Chapter 6, Following the Rhythms of TDD, demonstrates the red, green, refactor cycle as a rhythm of development. We decide on the next test to write, watch it fail, make it pass, and then refine our code to be safe and simple for the team to work with in the future.

Chapter 7, Driving Design – TDD and SOLID, builds on previous chapters showing how TDD provides rapid feedback on our design decisions by bringing SOLID into the fold. The SOLID principles are a useful set of guidelines to help design object-oriented code. This chapter is a review of those principles so that they are ready for us to apply in the rest of the book.

Chapter 8, Test Doubles – Stubs and Mocks, explains two critical techniques that allow us to swap out things that are difficult to test for things that are easier to test. By doing so, we can bring more of our code under a TDD unit test, reducing our need for slower integration testing.

Chapter 9, Hexagonal Architecture – Decoupling External Systems, presents a powerful design technique that enables us to fully decouple external systems such as databases and web servers from our core logic. We will introduce the concepts of ports and adapters here. This simplifies the use of TDD and as a benefit, provides resilience to any changes imposed on us by external factors.

Chapter 10, FIRST Tests and the Test Pyramid, outlines the test pyramid as a means of thinking about the different kinds of tests needed to fully test a software system. We discuss unit, integration, and end-to-end tests and the trade-offs between each type.

Chapter 11, How TDD Fits into Quality Assurance, explores how when using advanced test automation as described in this book, our QA engineers are freed up from some of the laborious detailed testing that they might have otherwise had to do. This chapter looks at how testing is now a whole-team effort throughout development, and how we can best combine our skills.

Chapter 12, Test First, Test Later, Test Never, reviews some different approaches to testing based on when we write the tests and what exactly we are testing. This will help us increase the quality of the tests we produce when applying TDD.

Chapter 13, Driving the Domain Layer, walks through applying TDD, SOLID, the test pyramid, and hexagonal architecture to the domain layer code of Wordz. Combined, these techniques enable us to bring most of the game logic under fast unit tests.

Chapter 14, Driving the Database Layer, offers guidance on writing the adapter code that connects to our SQL database, Postgres, now that we have decoupled the database code from the domain layer. We do this test-first, writing an integration test using the Database Rider test framework. The data access code is implemented using the JDBI library.

Chapter 15, Driving the Web Layer, rounds off as the final chapter of the book by explaining how to write an HTTP REST API that allows our Wordz game to be accessed as a web service. This is done test-first, using an integration test written using the tools built into the Molecule HTTP server library. Having completed this step, we will finally wire up the entire microservice ready to be run as a whole.

To get the most out of this book

This book assumes you know basic modern Java and can write short programs using classes, JDK 8 lambda expressions, and use the JDK 11 var keyword. It also assumes that you can use basic git commands, install software from web downloads onto your computer, and have basic familiarity with the IntelliJ IDEA Java IDE. Basic knowledge of SQL, HTTP, and REST will be helpful in the final chapters.

Software/hardware covered in the book	Operating system requirements
Amazon Corretto JDK 17 LTS	Windows, macOS, or Linux
IntelliJ IDEA 2022.1.3 Community Edition	Windows, macOS, or Linux
JUnit 5	Windows, macOS, or Linux
AssertJ	Windows, macOS, or Linux
Mockito	Windows, macOS, or Linux
DBRider	Windows, macOS, or Linux
Postgres	Windows, macOS, or Linux
psql	Windows, macOS, or Linux
Molecule	Windows, macOS, or Linux
git	Windows, macOS, or Linux

If you are using the digital version of this book, we advise you to type the code yourself or access the code from the book's GitHub repository (a link is available in the next section). Doing so will help you avoid any potential errors related to the copying and pasting of code.

Download the example code files

You can download the example code files for this book from GitHub at https://github.com/PacktPublishing/Test-Driven-Development-with-Java. If there's an update to the code, it will be updated in the GitHub repository.

We also have other code bundles from our rich catalog of books and videos available at https://github.com/PacktPublishing/. Check them out!

Download the color images

We also provide a PDF file that has color images of the screenshots and diagrams used in this book. You can download it here: https://packt.link/kLcmS.

Conventions used

There are a number of text conventions used throughout this book.

Code in text: Indicates code words in text, database table names, folder names, filenames, file extensions, pathnames, dummy URLs, user input, and Twitter handles. Here is an example: "Mount the downloaded WebStorm-10*.dmg disk image file as another disk in your system."

A block of code is set as follows:

```
public class DiceRoll {


    private final int NUMBER_OF_SIDES = 6;
    private final RandomGenerator rnd =
                        RandomGenerator.getDefault();
```

When we wish to draw your attention to a particular part of a code block, the relevant lines or items are set in bold:

```
public class DiceRoll {


    private final int NUMBER_OF_SIDES = 6;
    private final RandomGenerator rnd =
                        RandomGenerator.getDefault();
```

Any command-line input or output is written as follows:

```
private final int NUMBER_OF_SIDES = 6
```

Bold: Indicates a new term, an important word, or words that you see onscreen. For instance, words in menus or dialog boxes appear in **bold**. Here is an example: "Select **System info** from the **Administration** panel."

> Tips or important notes
> Appear like this.

Get in touch

Feedback from our readers is always welcome.

General feedback: If you have questions about any aspect of this book, email us at customercare@packtpub.com and mention the book title in the subject of your message.

Errata: Although we have taken every care to ensure the accuracy of our content, mistakes do happen. If you have found a mistake in this book, we would be grateful if you would report this to us. Please visit www.packtpub.com/support/errata and fill in the form.

Piracy: If you come across any illegal copies of our works in any form on the internet, we would be grateful if you would provide us with the location address or website name. Please contact us at copyright@packt.com with a link to the material.

If you are interested in becoming an author: If there is a topic that you have expertise in and you are interested in either writing or contributing to a book, please visit authors.packtpub.com.

Share Your Thoughts

Once you've read *Test-Driven Development with Java*, we'd love to hear your thoughts! Scan the QR code below to go straight to the Amazon review page for this book and share your feedback.

https://packt.link/r/1-803-23623-X

Your review is important to us and the tech community and will help us make sure we're delivering excellent quality content.

Download a free PDF copy of this book

Thanks for purchasing this book!

Do you like to read on the go but are unable to carry your print books everywhere? Is your eBook purchase not compatible with the device of your choice?

Don't worry, now with every Packt book you get a DRM-free PDF version of that book at no cost.

Read anywhere, any place, on any device. Search, copy, and paste code from your favorite technical books directly into your application.

The perks don't stop there, you can get exclusive access to discounts, newsletters, and great free content in your inbox daily

Follow these simple steps to get the benefits:

1. Scan the QR code or visit the link below

https://packt.link/free-ebook/978-1-80323-623-0

2. Submit your proof of purchase
3. That's it! We'll send your free PDF and other benefits to your email directly

Part 1:
How We Got to TDD

In *Part 1*, we look at how we got to TDD in the software industry. What problems are we trying to fix with TDD? What opportunities does it create?

In the following chapters, we will learn about the benefits that TDD brings to businesses and developers. We'll review the basics of good code to provide something to aim for when we first start writing our tests. Knowing that teams are sometimes hesitant to start using TDD, we will examine six common objections and how to overcome them.

This part has the following chapters:

- *Chapter 1, Building the Case for TDD*
- *Chapter 2, Using TDD to Create Good Code*
- *Chapter 3, Dispelling Common Myths about TDD*

1
Building the Case for TDD

Before we dive into what **test-driven development** (TDD) is and how to use it, we're going to need to understand why we need it. Every seasoned developer knows that bad code is easier to write than good code. Even good code seems to get worse over time. Why?

In this chapter, we will review the technical failures that make source code difficult to work with. We'll consider the effect that bad code has on both the team and the business bottom line. By the end of the chapter, we'll have a clear picture of the anti-patterns we need to avoid in our code.

In this chapter, we're going to cover the following main topics:

- Writing code badly
- Recognizing bad code
- Decreasing team performance
- Diminishing business outcomes

Writing code badly

As every developer knows, it seems a lot easier to write bad code than to engineer good code. We can define good code as being easy to understand and safe to change. Bad code is therefore the opposite of this, where it is very difficult to read the code and understand what problem it is supposed to be solving. We fear changing bad code – we know that we are likely to break something.

My own troubles with bad code go all the way back to my first program of note. This was a program written for a school competition, which aimed to assist realtors to help their customers find the perfect house. Written on the 8-bit Research Machines 380Z computer at school, this was 1981's answer to Rightmove.

In those pre-web days, it existed as a simple desktop application with a green-screen text-based user interface. It did not have to handle millions, never mind billions, of users. Nor did it have to handle millions of houses. It didn't even have a nice user interface.

As a piece of code, it was a couple of thousand lines of *Microsoft Disk BASIC 9* code. There was no code structure to speak of, just thousands of lines resplendent with uneven line numbers and festooned with global variables. To add an even greater element of challenge, BASIC limited every variable to a two-letter name. This made every name in the code utterly incomprehensible. The source code was intentionally written to have as few spaces in it as possible in order to save memory. When you only had 32KB of RAM to fit all of the program code, the data, and the operating system in, every byte mattered.

The program only offered its user basic features. The user interface was of its time, using only text-based forms. It predated graphical operating systems by a decade. The program also had to implement its own data storage system, using files on 5.25-inch floppy disks. Again, affordable database components were of the future. The main feature of the program in question was that users could search for houses within certain price ranges and feature sets. They could filter by terms such as the number of bedrooms or price range.

However, the code itself really was a mess. See for yourself – here is a photograph of the original listing:

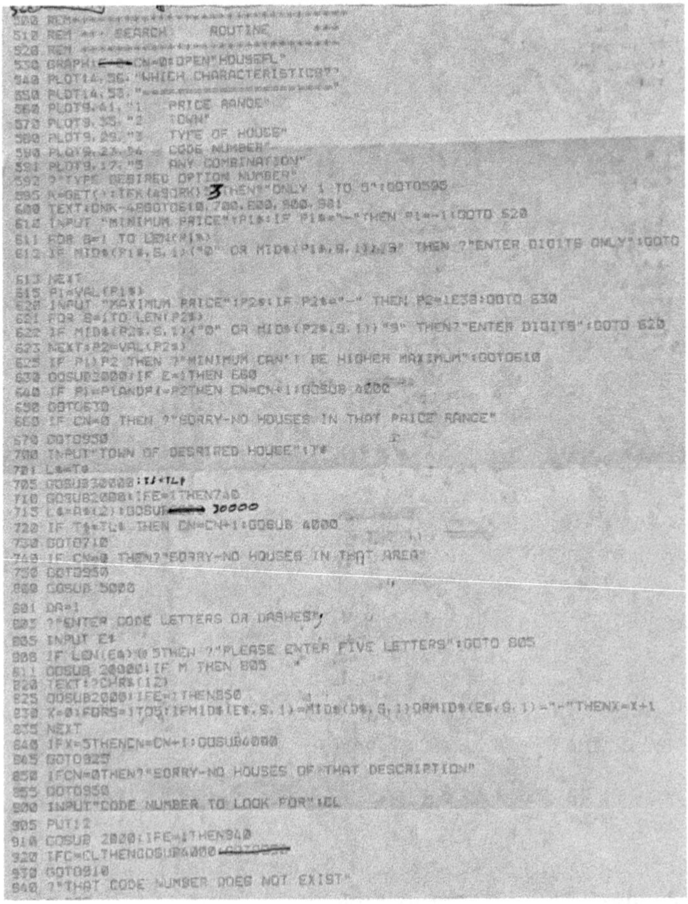

Figure 1.1 – The estate agent code listing

This horror is the original paper listing of one of the development versions. It is, as you can see, completely unreadable. It's not just you. Nobody would be able to read it easily. I can't and I wrote it. I would go as far as to say it is a mess, *my mess*, crafted by me, one keystroke at a time.

This kind of code is a nightmare to work with. It fails our definition of good code. It is not at all easy to read that listing and understand what the code is supposed to be doing. It is not safe to change that code. If we attempted to, we would find that we could never be certain about whether we have broken some feature or not. We would also have to manually retest the entire application. This would be time-consuming.

Speaking of testing, I never thoroughly tested that code. It was all manually tested without even following a formal test plan. At best, I would have run a handful of **happy path** manual tests. These were the kind of tests that would confirm that you could add or delete a house, and that some representative searches worked, but that was all. There was no way I ever tested every path through that code. I just guessed that it would work.

If the data handling had failed, I would not have known what had happened. I never tried it. Did every possible search combination work? Who knew? I certainly had no idea. I had even less patience to go through all that tedious manual testing. It worked, enough to win an award of sorts, but it was still bad code.

Understanding why bad code is written

In my case, it was simply down to a lack of knowledge. I did not know how to write good code. But there are also other reasons unrelated to skill. Nobody ever sets out to write bad code intentionally. Developers do the best job they can with the tools available and to the best of their ability at that time.

Even with the right skills, several common issues can result in bad code:

- A lack of time to refine the code due to project deadlines
- Working with legacy code whose structure prevents new code from being added cleanly
- Adding a short-term fix for an urgent production fault and then never reworking it
- Unfamiliarity with the subject area of the code
- Unfamiliarity with the local idioms and development styles
- Inappropriately using idioms from a different programming language

Now that we've seen an example of code that is difficult to work with, and understood how it came about, let's turn to the obvious next question: how can we recognize bad code?

Recognizing bad code

Admitting that our code is difficult to work with is one thing, but to move past that and write good code, we need to understand *why* code is bad. Let's identify the technical issues.

Bad variable names

Good code is self-describing and safe to change. Bad code is not.

Names are the most critical factor in deciding whether code will be easy to work with or not. Good names tell the reader clearly what to expect. Bad names do not. Variables should be named according to what they contain. They should answer *"why would I want to use this data? What will it tell me?"*

A string variable that has been named `string` is badly named. All we know is that it is a string. This does not tell us what is in the variable or why we would want to use it. If that string represented a surname, then by simply calling it `surname`, we would have helped future readers of our code understand our intentions much better. They would be able to easily see that this variable holds a surname and should not be used for any other purpose.

The two-letter variable names we saw in the listing in *Figure 1.1* represented a limitation of the BASIC language. It was not possible to do better at the time, but as we could see, they were not helpful. It is much harder to understand what `sn` means than `surname`, if that's what the variable stores. To carry that even further, if we decide to hold a surname in a variable named `x`, we have made things really difficult for readers of our code. They now have two problems to solve:

- They have to reverse-engineer the code to work out that `x` is used to hold a surname
- They have to mentally map `x` with the concept of surname every time that they use it

It is so much easier when we use descriptive names for all our data, such as local variables, method parameters, and object fields. In terms of more general guidelines, the following Google style guide is a good source: `https://google.github.io/styleguide/javaguide.html#s5-naming`.

> **Best practice for naming variables**
> Describe the data contained, not the data type.

We now have a better idea of how to go about naming variables. Now, let's look at how to name functions, methods, and classes properly.

Bad function, method, and class names

The names of functions, methods, and classes all follow a similar pattern. In good code, function names tell us why we should call that function. They describe what they will do for us as users of that function. The focus is on the outcome – what will have happened by the time the function returns.

We do not describe how that function is implemented. This is important. It allows us to change our implementation of that function later if that becomes advantageous, and the name will still describe the outcome clearly.

A function named `calculateTotalPrice` is clear about what it is going to do for us. It will calculate the total price. It won't have any surprising side effects. It won't try and do anything else. It will do what it says it will. If we abbreviate that name to `ctp`, then it becomes much less clear. If we call it `func1`, then it tells us absolutely nothing at all that is useful.

Bad names force us to reverse-engineer every decision made every time we read the code. We have to pore through the code to try and find out what it is used for. We should not have to do this. Names should be abstractions. A good name will speed up our ability to understand code by condensing a bigger-picture understanding into a few words.

You can think of the function name as a heading. The code inside the function is the body of text. It works just the same way that the text you're reading now has a heading, *Recognizing bad code*, which gives us a general idea of the content in the paragraphs that follow. From reading the heading, we expect the paragraphs to be about recognizing bad code, nothing more and nothing less.

We want to be able to skim-read our software through its *headings* – the function, method, class, and variable names – so that we can focus on what we want to do now, rather than relearning what was done in the past.

Method names are treated identically to function names. They both describe an action to be taken. Similarly, you can apply the same rules for function names to method names.

> **Best practice for method and function names**
> Describe the outcome, not the implementation.

Again, class names follow descriptive rules. A class often represents a single concept, so its name should describe that concept. If a class represents the user profile data in our system, then a class name of `UserProfile` will help readers of our code to understand that.

A name's length depends on namespacing

One further tip applies to all names with regard to their length. The name should be fully descriptive but its length depends on a few factors. We can choose shorter names when one of the following applies:

- The named variable has a small scope of only a few lines
- The class name itself provides the bulk of the description
- The name exists within some other namespace, such as a class name

Let's look at a code example for each case to make this clear.

The following code calculates the total of a list of values, using a short variable name, `total`:

```
int calculateTotal (List<Integer> values) {
    int total = 0;

    for ( Integer v : values ) {
        total += v;
    }

    return total ;
}
```

This works well because it is clear that `total` represents the total of all values. We do not need a name that is any longer given the context around it in the code. Perhaps an even better example lies in the v loop variable. It has a one-line scope, and within that scope, it is quite clear that v represents the current value within the loop. We could use a longer name such as `currentValue` instead. However, does this add any clarity? Not really.

In the following method, we have a parameter with the short name `gc`:

```
private void draw (GraphicsContext gc) {
    // code using gc omitted
}
```

The reason we can choose such a short name is that the `GraphicsContext` class carries most of the description already. If this were a more general-purpose class, such as `String`, for example, then this short name technique would be unhelpful.

In this final code example, we are using the short method name of `draw()`:

```
public class ProfileImage {
    public void draw (WebResponse wr) {
        // Code omitted
    }
}
```

The class name here is highly descriptive. The `ProfileImage` class name we've used in our system is one that is commonly used to describe the avatar or photograph that shows on a user's profile page. The `draw()` method is responsible for writing the image data to a `WebResponse` object. We could choose a longer method name, such as `drawProfileImage()`, but that simply repeats information that has already been made clear given the name of the class. Details such as this are what give Java its

reputation for being verbose, which I feel is unfair; it is often us Java programmers who are verbose, rather than Java itself.

We've seen how properly naming things makes our code easier to understand. Let's take a look at the next big problem that we see in bad code – using constructs that make logic errors more likely.

Error-prone constructs

Another tell-tale sign of bad code is that it uses error-prone constructs and designs. There are always several ways of doing the same thing in code. Some of them provide more scope to introduce mistakes than others. It therefore makes sense to choose ways of coding that actively avoid errors.

Let's compare two different versions of a function to calculate a total value and analyze where errors might creep in:

```java
int calculateTotal(List<Integer> values) {
    int total = 0;

    for ( int i=0; i<values.size(); i++) {
        total += values.get(i);
    }

    return total ;
}
```

The previous listing is a simple method that will take a list of integers and return their total. It's the sort of code that has been around since **Java 1.0.2**. It works, yet it is error prone. In order for this code to be correct, we need to get several things right:

- Making sure that `total` is initialized to 0 and not some other value
- Making sure that our `i` loop index is initialized to 0
- Making sure that we use < and not <= or == in our loop comparison
- Making sure that we increment the `i` loop index by exactly one
- Making sure that we add the value from the current index in the list to `total`

Experienced programmers do tend to get all this right first time. My point is that there is a *possibility* of getting any or all of these things wrong. I've seen mistakes made where <= has been used instead of < and the code fails with an `ArrayIndexOutOfBounds` exception as a result. Another easy mistake is to use = in the line that adds to the total value instead of +=. This has the effect of returning only the last value, not the total. I have even made that mistake as a pure *typo* – I honestly thought I had typed the right thing but I was typing quickly and I hadn't.

It is clearly much better for us to *avoid* these kinds of errors entirely. If an error cannot happen, then it will not happen. This is a process I call *designing out errors*. It is a fundamental clean-code practice. To see how we could do this to our previous example, let's look at the following code:

```
int calculateTotal(List<Integer> values) {
    return values.stream().mapToInt(v -> v).sum();
}
```

This code does the same thing, yet it is inherently safer. We have no `total` variable, so we cannot initialize that incorrectly, nor can we forget to add values to it. We have no loop and so no loop index variable. We cannot use the wrong comparison for the loop end and so cannot get an `ArrayIndexOutOfBounds` exception. There is simply far less that can go wrong in this implementation of the code. It generally makes the code clearer to read as well. This, in turn, helps with onboarding new developers, code reviews, adding new features, and pair programming.

Whenever we have a choice to use code with fewer parts that could go wrong, we should choose that approach. We can make life easier for ourselves and our colleagues by choosing to keep our code as error-free and simple as possible. We can use more robust constructs to give bugs fewer places to hide.

It is worth mentioning that both versions of the code have an integer overflow bug. If we add integers together whose total is beyond the allowable range of -2147483648 to 2147483647, then the code will produce the wrong result. The point still stands, however: the later version has fewer places where things can go wrong. *Structurally*, it is simpler code.

Now that we have seen how to avoid the kinds of errors that are typical of bad code, let's turn to other problem areas: coupling and cohesion.

Coupling and cohesion

If we have a number of Java classes, **coupling** describes the relationship between those classes, while **cohesion** describes the relationships between the methods inside each one.

Our software designs become easier to work with once we get the amounts of coupling and cohesion right. We will learn techniques to help us do this in *Chapter 7, Driving Design–TDD and SOLID*. For now, let's understand the problems that we will face when we get this wrong, starting with the problem of low cohesion.

Low cohesion inside a class

Low cohesion describes code that has many different ideas all lumped together in it in a single place. The following UML class diagram shows an example of a class with low cohesion among its methods:

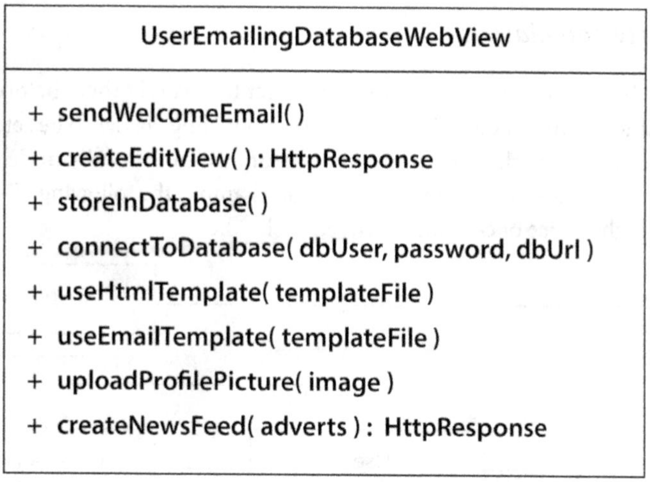

UserEmailingDatabaseWebView
+ sendWelcomeEmail( )
+ createEditView( ) : HttpResponse
+ storeInDatabase( )
+ connectToDatabase(dbUser, password, dbUrl)
+ useHtmlTemplate(templateFile)
+ useEmailTemplate(templateFile)
+ uploadProfilePicture(image)
+ createNewsFeed(adverts) : HttpResponse

Figure 1.2 – Low cohesion

The code in this class attempts to combine too many responsibilities. They are not all obviously related – we are writing to a database, sending out welcome emails, and rendering web pages. This large variety of responsibilities makes our class harder to understand and harder to change. Consider the different reasons we may need to change this class:

- Changes to the database technology

- Changes to the web view layout

- Changes to the web template engine technology

- Changes to the email template engine technology

- Changes to the news feed generation algorithm

There are many reasons why we would need to change the code in this class. It is always better to give classes a more precise focus, so that there are fewer reasons to change them. Ideally, any given piece of code should only have one reason to be changed.

Understanding code with low cohesion is hard. We are forced to understand many different ideas at once. Internally, the code is very interconnected. Changing one method often forces a change in others because of this. Using the class is difficult, as we need to construct it with all its dependencies. In our example, we have a mixture of templating engines, a database, and code for creating a web page. This also makes the class very difficult to test. We need to set up all these things before we can run test methods against that class. Reuse is limited with a class like this. The class is very tightly bound to the mix of features that are rolled into it.

High coupling between classes

High coupling describes where one class needs to connect to several others before it can be used. This makes it difficult to use in isolation. We need those supporting classes to be set up and working correctly before we can use our class. For the same reason, we cannot fully understand that class without understanding the many interactions it has. As an example, the following UML class diagram shows classes with a high degree of coupling between each other:

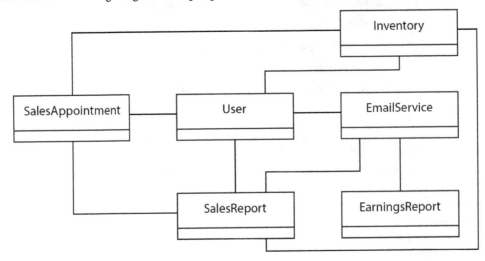

Figure 1.3 – High coupling

In this fictitious example of a sales tracking system, several of the classes need to interact with each other. The User class in the middle couples to four other classes: Inventory, EmailService, SalesAppointment, and SalesReport. This makes it harder to use and test than a class that couples to fewer other classes. Is the coupling here too high? Maybe not, but we can imagine other designs that would reduce it. The main thing is to be aware of the degree of coupling that classes have in our designs. As soon as we spot classes with many connections to others, we know we are going to have a problem understanding, maintaining, and testing them.

We've seen how the technical elements of high coupling and low cohesion make our code difficult to work with, but there is a social aspect to bad code as well. Let's consider the effect bad code has on the development team.

Decreasing team performance

A good way to look at bad code is code lacking the technical practices that help other developers understand what it is doing.

When you're coding solo, it doesn't matter so much. Bad code will just slow you down and feel a little demoralizing at times. It does not affect anybody else. However, most professionals code in development teams, which is a whole different ball game. Bad code really slows a team down.

The following two studies are interesting as far as this is concerned:

- `https://dl.acm.org/doi/abs/10.1145/3194164.3194178`
- `https://www.sciencedirect.com/science/article/abs/pii/S0164121219301335`

The first study shows that developers waste up to 23% of their time on bad code. The second study shows that in 25% of cases of working with bad code, developers are forced to increase the amount of bad code still further. In these two studies, the term **technical debt** is used, rather than referring to bad code. There is a difference in intention between the two terms. Technical debt is code that is shipped with known technical deficiencies in order to meet a deadline. It is tracked and managed with the intention that it will later be replaced. Bad code might have the same defects, but it lacks the redeeming quality of intentionality.

It is all too easy to check in code that has been easy to write but will be hard to read. When I do that, I have effectively placed a tax on the team. The next developer to pull my changes will have to figure out what on earth they need to do and my bad code will have made that much harder.

We've all been there. We start a piece of work, download the latest code, and then just stare at our screens for ages. We see variable names that make no sense, mixed up with tangled code that really does not explain itself very well at all. It's frustrating for us personally, but it has a real cost in a programming business. Every minute we spend not understanding code is a minute where money is being spent on us achieving nothing. It's not what we dreamed of when we signed up to be a developer.

Bad code disrupts every future developer who has to read the code, even us, the original authors. We forget what we previously meant. Bad code means more time spent by developers fixing mistakes, instead of adding value. It means more time is lost on fixing bugs in production that should have been easily preventable.

Worse still, this problem compounds. It is like interest on a bank loan. If we leave bad code in place, the next feature will involve adding workarounds for the bad code. You may see extra conditionals appear, giving the code yet more execution paths and creating more places for bugs to hide. Future features build on top of the original bad code and all of its workarounds. It creates code where most of what we read is simply working around what never worked well in the first place.

Code of this kind drains the motivation out of developers. The team starts spending more time working around problems than they spend adding value to the code. None of this is *fun* for the typical developer. It's not fun for anybody on the team.

Project managers lose track of the project status. Stakeholders lose confidence in the team's ability to deliver. Costs overrun. Deadlines slip. Features get quietly cut, just to claw back a little slack in the schedule. Onboarding new developers becomes painful, to the point of awkwardness, whenever they see the awful code.

Bad code leaves the whole team unable to perform to the level they are capable of. This, in turn, does not make for a happy development team. Beyond unhappy developers, it also negatively impacts business outcomes. Let's understand those consequences.

Diminishing business outcomes

It's not just the development team who suffers from the effects of bad code. It's bad for the entire business.

Our poor users end up paying for software that doesn't work, or at least that doesn't work properly. There are many ways that bad code can mess up a user's day, whether as a result of lost data, unresponsive user interfaces, or any kind of intermittent fault. Each one of these can be caused by something as trivial as setting a variable at the wrong time or an off-by-one error in a conditional somewhere.

The users see neither any of that nor the thousands of lines of code that we got right. They just see their missed payment, their lost document that took 2 hours to type, or that fantastic *last-chance* ticket deal that simply never happened. Users have little patience for things like this. Defects of this kind can easily lose us a valuable customer.

If we are lucky, users will fill out a bug report. If we are really lucky, they will let us know what they were doing at the time and provide us with the right steps to reproduce the fault. But most users will just hit delete on our app. They'll cancel future subscriptions and ask for refunds. They'll go to review sites and let the world know just how useless our app and company are.

At this point, it isn't merely bad code; it is a commercial liability. The failures and honest human errors in our code base are long forgotten. Instead, we were just a competitor business that came and went in a blaze of negativity.

Decreased revenue leads to decreased market share, a reduced **Net Promoter Score**®™ (**NPS**), disappointed shareholders, and all the other things that make your C-suite lose sleep at night. Our bad code has become a problem at the business level.

This isn't hypothetical. There have been several incidents where software failures have cost the business. Security breaches for Equifax, Target, and even the Ashley Madison site all resulted in losses. The Ariane rocket resulted in the loss of both spacecraft and satellite payload, a total cost of billions of dollars! Even minor incidents resulting in downtime for e-commerce systems can soon have costs mounting, while consumer trust crashes down.

In each case, the failures may have been small errors in comparatively few lines of code. Certainly, they will have been avoidable in some way. We know that humans make mistakes, and that all software is built by humans, yet a little extra help may have been all it would have taken to stop these disasters from unfolding.

The advantage of finding failures early is shown in the following diagram:

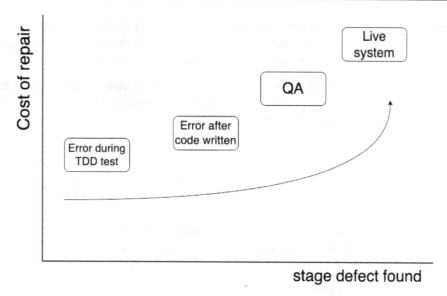

Figure 1.4 – Costs of defect discovery

In the previous figure, the cost of the repair of a defect gets higher the later it is found:

- Found by a failing test before code:

  The cheapest and fastest way to discover a defect is by writing a test for a feature before we write the production code. If we write the production code that we expect should make the test pass, but instead the test fails, we know there is a problem in our code.

- Found by a failing test after code:

  If we write the production code for a feature, and then write a test afterward, we may find defects in our production code. This happens a little later in the development cycle. We will have wasted a little more time before discovering the defect.

- Found during manual QA:

  Many teams include **Quality Assurance (QA)** engineers. After code has been written by a developer, the QA engineer will manually test the code. If a defect is found here, this means significant time has passed since the developer first wrote the code. Rework will have to be done.

- Found by the end user once code is in production:

  This is as bad as it gets. The code has been shipped to production and end users are using it. An end user finds a bug. The bug has to be reported, triaged, a fix scheduled for development, then retested by QA then redeployed to production. This is the slowest and most expensive path to discovering a defect.

The earlier we find the fault, the less time and money we will have to spend on correcting it. The ideal is to have a failing test before we even write a line of code. This approach also helps us design our code. The later we leave it to find a mistake, the more trouble it causes for everyone.

We've seen how low-quality code gives rise to defects and is bad for business. The earlier we detect failures, the better it is for us. Leaving defects in production code is both difficult and expensive to fix, and negatively affects our business reputation.

Summary

We can now recognize bad code from its technical signs and appreciate the problems that it causes for both development teams and business outcomes.

What we need is a technique to help us avoid these problems. In the next chapter, we'll take a look at how TDD helps us deliver clean, correct code that is a true business asset.

Questions and answers

1. Isn't it enough to have working code?

    Sadly not. Code that meets user needs is an entry-level step with professional software. We also need code that we *know* works, and that the team can easily understand and modify.

2. Users don't see the code. Why does it matter to them?

    This is true. However, users expect things to work reliably, and they expect our software to be updated and improved continuously. This is only possible when the developers can work safely with the existing code.

3. Is it easier to write good code or bad code?

    It is much harder to write good code, unfortunately. Good code does more than simply work correctly. It must also be easy to read, easy to change, and safe for our colleagues to work with. That's why techniques such as TDD have an important role to play. We need all the help we can get to write clean code that helps our colleagues.

Further reading

- More about the loss of the Ariane rocket: `https://www.esa.int/Newsroom/Press_Releases/Ariane_501_-_Presentation_of_Inquiry_Board_report`

2
Using TDD to Create Good Code

We've seen that bad code is bad news: bad for business, bad for users, and bad for developers. **Test-driven development (TDD)** is a core software engineering practice that helps us keep bad code out of our systems.

The goal of this chapter is to learn the specifics of how TDD helps us to create well-engineered, correct code, and how it helps us to keep it that way. By the end, we will understand the basic principles behind good code and how TDD helps us create it. It is important for us to understand why TDD works in order to motivate us and so that we have a response to give to colleagues about why we recommend that they use it as well.

In this chapter, we're going to cover the following main topics:

- Designing good quality code
- Revealing design flaws
- Preventing logic flaws
- Protecting against future defects
- Documenting our code

Designing good quality code

Good quality code doesn't happen by accident. It is intentional. It is the result of thousands of small decisions, each one shaping how easy our code is to read, test, compose, and change. We must choose between quick-and-dirty hacks, where we have no idea what edge cases are covered, and more robust approaches, where we are confident that no matter how the user misuses our code, it will work as expected.

Every line of source code involves at least one of these decisions. That's an awful lot of deciding that we have to do.

You'll notice that we haven't mentioned TDD so far. As we will see, TDD does not design your code for you. It doesn't remove that essential engineering sensibility and creative input that is needed to turn requirements into code. To be honest, I'm grateful for that – it's the part that I enjoy.

However, that does cause a lot of early failure with TDD, which is worth noting. Expecting to *implement the TDD process* and get good quality code out without your own design input will simply not work. TDD, as we will see, is a tool that allows you to get rapid feedback on these design decisions. You can change your mind and adapt while the code is still cheap and quick to change but they are still *your* design decisions that are playing out.

So, what is good code? What are we aiming for?

Good code, for me, is all about readability. I optimize for clarity. I want to be kind to my future self and my long-suffering colleagues by engineering code that is clear and safe to work with. I want to create clear and simple code that is free of hidden traps.

While there is a huge range of advice on what makes good code, the basics are straightforward:

- Say what you mean, mean what you say

- Take care of the details in private

- Avoid accidental complexity

It's worth a quick review of what I mean by those things.

Say what you mean, mean what you say

Here's an interesting experiment. Take a piece of source code (in any language) and strip out everything that is not part of the language specification, then see if you can figure out what it does. To make things really stand out, we will replace all method names and variable identifiers with the symbol ???.

Here's a quick example:

```
public boolean ??? (int ???) {
    if ( ??? > ??? ) {
        return ???;
    }

    return ???;
}
```

Any ideas what this code does? No, me neither. I haven't a clue.

I can tell by its shape that it is some kind of *assessment* method that passes something in and returns `true`/`false`. Maybe it implements a threshold or limit. It uses a multipath return structure, where we check something, then return an answer as soon as we know what that answer is.

While the shape of the code and the syntax tell us something, it's not telling us much. It is definitely not enough. Nearly all the information we share about what our code does is a result of the natural language identifiers we choose. Names are absolutely vital to good code. They are beyond important. They are everything. They can reveal intent, explain outcomes, and describe why a piece of data is important to us, but they can't do any of this if we do a bad job choosing our names.

I use two guidelines for names, one for naming active code – methods and functions – and one for variables:

- **Method** – Say what it does. What is the outcome? Why would I call this?
- **Variable** – Say what it contains. Why would I access this?

A common mistake with method naming is to describe how it works internally, instead of describing what the outcome is. A method called `addTodoItemToItemQueue` is committing us to one specific implementation of a method that we don't really care about. Either that or it is misinformation. We can improve the name by calling it `add(Todo item)`. This name tells us why exactly we should call this method. It leaves us free to revise how it is coded later.

The classic mistake with variable names is to say what they are made of. For example, the variable name `String string` helps nobody, whereas `String firstName` tells me clearly that this variable is somebody's first name. It tells me why I would want to read or write that variable.

Perhaps more importantly, it tells us what *not* to write in that variable. Having one variable serve multiple purposes in the same scope is a real headache. Been there, done that, never going back.

It turns out that code is *storytelling*, pure and simple. We tell the story of what problem we are solving and how we have decided to solve it to human programmers. We can throw any old code into a compiler and the computer will make it work but we must take more care if we want humans to understand our work.

Take care of the details in private

Taking care of the details in private is a simple way to describe the computer science concepts of **abstraction** and **information hiding**. These are fundamental ideas that allow us to break complex systems into smaller, simpler parts.

The way I think about abstraction is the same way I think about hiring an electrician for my house.

I know that my electric water heater needs to be fixed but I don't want to know how. I don't want to learn how to do it. I don't want to have to figure out what tools are needed and buy them. I want to have nothing whatsoever to do with it, beyond asking that it gets done when I need it done. So, I'll call the electrician and ask them to do it. I'm more than happy to pay for a good job, as long as I don't have to do it myself.

This is what abstraction means. The electrician abstracts the job of fixing my water heater. Complex stuff gets done in response to my simple requests.

Abstraction happens everywhere in good software.

Every time you make some kind of detail less important, you have abstracted it. A method has a simple signature but the code inside it may be complex. This is an abstraction of an algorithm. A local variable might be declared as type `String`. This is an abstraction of the memory management of each text character and the character encoding. A microservice that will send discount vouchers to our top customers who haven't visited the site in a while is an abstraction of a business process. Abstraction is everywhere in programming, across all major paradigms – **object-oriented programming** (**OOP**), **procedural**, and **functional**.

The idea of splitting software into components, each of which takes care of something for us, is a massive quality driver. We centralize decisions, meaning that we don't make mistakes in duplicated code. We can test a component thoroughly in isolation. We *design out* problems caused by hard-to-write code just by writing it once and having an easy-to-use interface.

Avoid accidental complexity

This is my personal favorite destroyer of good code – complex code that simply never needed to exist.

There are always many ways of writing a piece of code. Some of them use complicated features or go all around the houses; they use convoluted chains of actions to do a simple thing. All versions of the code get the same result but some just do it in a more complicated way by accident.

My goal for code is to tell at first sight the story of what problem I am solving, leaving the details about how I am solving it for closer analysis. This is quite different from how I learned how to code originally. I choose to emphasize **domain** over **mechanism**. The domain here means using the same language as the user, for example, expressing the problem in business terms, not just raw computer code syntax. If I am writing a banking system, I want to see money, ledgers, and transactions coming to the forefront. The story the code is telling has to be that of banking.

Implementation details such as message queues and databases are important but only as far as they describe how we are solving the problem today. They may need to change later. Whether they change or not, we still want the primary story to be about *transactions going into an account* and not *message queues talking to REST services*.

As our code gets better at telling the story of the problem we are solving, we make it easier to write replacement components. Swapping out a database for another vendor's product is simplified because we know exactly what purpose it is serving in our system.

This is what we mean by hiding details. At some level, it is important to see how we wired up the database, but only after we have seen why we even needed one in the first place.

To give you a concrete example, here is a piece of code similar to some code that I found in a production system:

```java
public boolean isTrue (Boolean b) {
    boolean result = false;

    if ( b == null ) {
        result = false;
    }
    else if ( b.equals(Boolean.TRUE)) {
        result = true;
    }
    else if ( b.equals(Boolean.FALSE)) {
        result = false;
    }
    else {
        result = false;
    }

    return result;
}
```

You can see the problem here. Yes, there is a need for a method like this. It is a low-level mechanism that converts a Java true/false object into its equivalent primitive type and does it safely. It covers all edge cases relating to a null value input, as well as valid true/false values.

However, it has problems. This code is cluttered. It is unnecessarily hard to read and test. It has high **cyclomatic complexity (CYC)**. CYC is an objective measure of how complex a piece of code is, based on the number of independent execution paths possible in a section of code.

The previous code is unnecessarily verbose and over-complicated. I'm pretty sure it has a **dead-code path** – meaning a path containing unreachable code – on that final else, as well.

Looking at the logic needed, there are only three interesting input conditions: null, true, and false. It certainly does not need all those else/if chains to decode that. Once you've got that *null-to-false* conversion out of the way, you really only need to inspect one value before you can fully decide what to return.

A better equivalent would be the following:

```
public boolean isTrue (Boolean b) {
    return Boolean.TRUE.equals(b);
}
```

This code does the same thing with a lot less fuss. It does not have the same level of accidental complexity as the previous code. It reads better. It is easier to test with fewer paths needing testing. It has a better cyclomatic complexity figure, which means fewer places for bugs to hide. It tells a better story about why the method exists. To be perfectly honest, I might even refactor this method by inlining it. I'm not sure the method adds any worthwhile extra explanation to the implementation.

This method was a simple example. Just imagine seeing this scaled up to thousands of lines of copy-pasted, slightly-changed code. You can see why accidental complexity is a killer. This cruft builds up over time and grows exponentially. Everything becomes harder to read and harder to safely change.

Yes, I have seen that. I will never stop being sad about it when I do. We can do better than this. As professional software engineers, we really should.

This section has been a lightning tour of *good design* fundamentals. They apply across all styles of programming. However, if we can do things right, we can also do things wrong. In the next section, we'll take a look at how TDD tests can help us prevent bad designs.

Revealing design flaws

Bad design is truly bad. It is the root cause of software being hard to change and hard to work with. You can never quite be sure whether your changes are going to work because you can never quite be sure what a bad design is really doing. Changing that kind of code is scary and often gets put off. Whole sections of code can be left to rot with only a `/* Here be dragons! */` comment to show for it.

The first major benefit of TDD is that it forces us to think about the design of a component. We do that before we think about how we implement it. By doing things in this order, we are far less likely to drift into a bad design by mistake.

The way we consider the design first is to think about the public interfaces of a component. We think about how that component will be used and how it will be called. We don't yet consider how we will make any implementations actually work. This is *outside-in* thinking. We consider the usage of the code from outside callers before we consider any inside implementation.

This is quite a different approach to take for many of us. Typically, when we need code to do something, we start by writing the implementation. After that, we will ripple out whatever is needed in method signatures, without a thought about the call site. This is *inside-out* thinking. It works, of course, but it often leads to complex calling code. It locks us into implementation details that just aren't important.

Outside-in thinking means we get to dream up the perfect component for its users. Then, we will bend the implementation to work with our desired code at the call site. Ultimately, this is far more important than the implementation. This is, of course, abstraction being used in practice.

We can ask questions like the following:

- Is it easy to set up?

- Is it easy to ask it to do something?

- Is the outcome easy to work with?

- Is it difficult to use it the wrong way?

- Have we made any incorrect assumptions about it?

You can see that by asking the right sort of questions, we're going to get the right sort of results.

By writing tests first, we cover all these questions. We decide upfront how we are going to set up our component, perhaps deciding on a clear constructor signature for an object. We decide how we are going to make the calling code look and what the call site will be. We decide how we will consume any results returned or what the effect will be on collaborating components.

This is the heart of software design. TDD does not do this for us, nor does it force us to do a good job. We could still come up with terrible answers for all those questions and simply write a test to lock those poor answers into place. I've seen that happen on numerous occasions in real code as well.

TDD provides that early opportunity to reflect on our decisions. We are literally writing the first example of a working, executable call site for our code before we even think about how it will work. We are totally focused on how this new component is going to fit into the bigger picture.

The test itself provides immediate feedback on how well our decisions have worked out. It gives three tell-tale signals that we could and should improve. We'll save the details for a later chapter but the test code itself clearly shows when your component is either hard to set up, hard to call, or its outputs are hard to work with.

Analyzing the benefits of writing tests before production code

There are three times you can choose to write tests: before the code, after the code, or never.

Obviously, never writing any tests sends us back to the dark ages of development. We're winging it. We write code assuming it will work, then leave it all to a manual test stage later. If we're lucky, we will discover functional errors at this stage, before our customers do.

Writing tests just after we complete a small chunk of code is a much better option. We get much faster feedback. Our code isn't necessarily any better though, because we write with the same mindset as we do without the implementation of tests. The same kinds of functional errors will be present. The good news is that we will then write tests to uncover them.

This is a big improvement, but it still isn't the gold standard, as it leads to a couple of subtle problems:

- Missing tests
- Leaky abstractions

Missing tests – undetected errors

Missing tests happen because of human nature. When we are busy writing code, we are juggling many ideas in our heads at once. We focus on specific details at the expense of others. I always find that I mentally *move on* a bit too quickly after a line of code. I just assume that it's going to be okay. Unfortunately, when I come to write my tests, that means I've forgotten some key points.

Suppose I end up writing some code like this:

```
public boolean isAllowed18PlusProducts( Integer age ) {
    return (age != null)  && age.intValue() > 18;
}
```

I'll probably have quickly started with the > 18 check, then *moved on* mentally and remembered that the age could be null. I will have added the And clause to check whether it is or not. That makes sense. My experience tells me that this particular snippet of code needs to do more than be a basic, robust check.

When I write my test, I'll remember to write a test for what happens when I pass in null, as that is fresh in my mind. Then, I will write another test for what happens with a higher age, say *21*. Again, good.

Chances are that I will forget about writing a test for the edge case of an age value of 18. That's really important here but my mind has moved on from that detail already. All it will take is one Slack message from a colleague about what's for lunch, and I will most likely forget all about that test and start coding the next method.

The preceding code has a subtle bug in it. It is supposed to return true for any age that is 18 or above. It doesn't. It returns true only for 19 and above. The greater-than symbol should have been a greater-than-or-equal-to symbol but I missed this detail.

Not only did I miss the nuance in the code but I missed out a vital test. I wrote two important tests but I needed three.

Because I wrote the other tests, I get no warning at all about this. You don't get a failing test that you haven't written.

We can avoid this by writing a failing test for every piece of code, then adding only enough code to make that test pass. That workflow would have been more likely to steer us toward thinking through the four tests needed to drive out null handling and the three boundary cases relating to age. It cannot guarantee it, of course, but it can drive the right kind of thinking.

Leaky abstractions – exposing irrelevant details

Leaky abstractions are a different problem. This is where we focus so much on the inside of the method that we forget to think about our *dream call site*. We just ripple out whatever is easiest to code.

We might be writing an interface where we store UserProfile objects. We might proceed code-first, pick ourselves a JDBC library that we like, code up the method, then find that it needs a database connection.

We might simply add a Connection parameter to fix that:

```
interface StoredUserProfiles {
    UserProfile load( Connection conn, int userId );
}
```

At first sight, there's nothing much wrong with it. However, look at that first parameter: it's the JDBC-specific Connection object. We have locked our interface into having to use JDBC. Or at the very least, having to supply some JDBC-related thing as a first parameter. We didn't even mean to do that. We simply hadn't thought about it thoroughly.

If we think about the ideal abstraction, it should load the corresponding UserProfile object for the given userId. It should *not* know how it is stored. The JDBC-specific Connection parameter should not be there.

If we think outside-in and consider the design before the implementation, we are less likely to go down this route.

Leaky abstractions like this create accidental complexity. They make code harder to understand by forcing future readers to wonder why we are insisting on JDBC use when we never meant to do so. We just forgot to design it out.

Writing tests first helps prevent this. It leads us to think about the ideal abstractions as a first step so we can write the test for them.

Once we have that test coded up, we have locked in our decision on how the code will be used. Then, we can figure out how to implement that without any unwanted details leaking out.

The previously explained techniques are simple but cover most of the basics of good design. Use clear names. Use simple logic. Use abstraction to hide implementation details, so that we emphasize what problem we are solving, rather than how we are solving it. In the next section, let's review the most obvious benefit of TDD: preventing flaws in our logic.

Preventing logic flaws

The idea of logic errors is perhaps what everybody thinks of first when we talk about testing: *did it work right?*

I can't disagree here – this is really important. As far as users, revenues, our Net Promoter Score®™, and market growth go, if your code doesn't work right, it doesn't sell. It's that simple.

Understanding the limits of manual testing

We know from bitter experience that the simplest logic flaws are often the easiest to create. The examples that we can all relate to are those one-off errors, that `NullPointerException` from an uninitialized variable, and that exception thrown by a library that wasn't in the documentation. They are all so simple and small. It seems like it would be so obvious for us to realize that we were making these mistakes, yet we all know they are often the hardest to spot. When we humans concentrate on the big picture of our code, sometimes these critical details just go unnoticed.

We know that manual testing *can* reveal these logic flaws but we also know from experience that manual test plans are fragile. It is possible to miss steps out or rush and miss important errors. We might simply assume that something does not need testing on this release *because we did not change that section of code*. You guessed it – that doesn't always work out so well for us. Bugs can arise in sections of code that seem totally unrelated to the bug if some underlying assumption has changed.

Manual testing costs money, which is money that can now not be spent on adding shiny new features instead.

Manual testing also gets blamed for delaying ship dates. Now, this is spectacularly unfair to our manual test colleagues. The development team – obviously writing code without TDD tests – stumble over their own bugs until there are only a couple of days left to ship. Then, we hand over the code to the testers, who have to run a huge test document in next to no time. They sometimes get blamed for *delaying the release*, even though the real cause was development taking longer than it should.

Yet, we never truly had a release. If we define a release as including tested code, which we should, then it is clear that the necessary testing never happened. You can't ethically release code when you don't even know whether it works. If you do, your users will be quick to complain.

It's no wonder some of my testing colleagues get so grumpy by the end of a sprint.

Solving problems by automating the tests

TDD has this totally covered. These logic errors simply cannot arise, which sounds like fantasy, but it really is true.

Before you type any production code, you have already written a failing test. Once you add your new code, you rerun the test. If you somehow typed in a logic error, the test still fails and you know about it *right away*. That's the magic here: your mistake happens but is highlighted right away. This enables you to fix it when it is fresh in your mind. It also means you cannot forget about fixing it later on.

You can often go to the exact line that's wrong and make the change. It's 10 seconds of work, not months of waiting for a test silo to get to work and fill out a *JIRA* bug ticket.

The kinds of unit tests we are talking about are also fast to run – very fast. Many of them run within a millisecond. Compare that to the total time to write a test plan document, run the whole app, set up stored data, operate the **user interface** (**UI**), record output, then write up a bug ticket. It is incomparably better, isn't it?

You can see how this is a bug-squashing superpower. We are making significant time savings within the code-test-debug cycle. This reduces development costs and increases delivery velocity. These are big wins for our team and our users.

Every time you write a test before code, you have kept bugs out of that code. You follow the most basic rule that you do not check code with failing tests. You make them pass.

It shouldn't need saying but you also don't cheat around that failing test by deleting it, ignoring it, or making it *always pass* by using some technical hack. However, I am saying all this because I have seen exactly that done in real code.

We've seen how writing tests first helps prevent adding bugs in our new code but TDD is even better than that: it helps prevent adding bugs in code that we will add *in the future*, which we will cover in the next section.

Protecting against future defects

As we grow our code by writing tests first, we could always simply delete each test after it has passed. I've seen some students do that when I've taught them TDD because I hadn't explained that we shouldn't do that yet. Regardless, we don't delete tests once they pass. We keep them all.

Tests grow into large regression suites, automatically testing every feature of the code we have built. By frequently running all the tests, we gain safety and confidence in the entire code base.

As team members add features to this code base, keeping all the tests passing shows that nobody has accidentally broken something. It is quite possible in software to add a perfectly innocent change somewhere, only to find that some seemingly unrelated thing has now stopped working. This will be because of the relationship between those two pieces that we previously did not understand.

The tests have now caused us to learn more about our system and our assumptions. They have prevented a defect from being written into the code base. These are both great benefits but the bigger picture is that our team has the confidence to make changes safely and know they have tests automatically looking after them.

This is true agility, the freedom to change. Agility was never about JIRA tickets and sprints. It was always about the ability to move quickly, with confidence, through an ever-changing landscape of requirements. Having tens of thousands of fast-running automated tests is probably the biggest enabling practice we have.

The ability of tests to give team members confidence to work quickly and effectively is a huge benefit of TDD. You may have heard the phrase *move fast and break things*, famous from the early days of Facebook. TDD allows us to move fast and *not* break things.

As we've seen, tests are great at providing fast feedback on design and logic correctness, as well as providing a defense against future bugs, but one huge extra benefit is that tests document our code.

Documenting our code

Everybody likes helpful, clear documentation, but not when it is out of date and unrelated to the current code base.

There is a general principle in software that the more separation there is between two related ideas, the more pain they will bring. As an example, think of some code that reads some obscure file format that nobody remembers. All works well, so long as you are reading files in that old format. Then you upgrade the application, that old file format is no longer supported, and everything breaks. The code was separated from the data content in those old files. The files didn't change but the code did. We didn't even realize what was going on.

It's the same with documentation. The worst documentation is often contained in the glossiest productions. These are artifacts written a long time after the code was created by teams with separate skillsets – copywriting, graphic design, and so on. Documentation updates are the first thing to get dropped from a release when time gets tight.

The solution is to bring documentation closer to the code. Get it produced by people closer to the code who know how it works in detail. Get it read by people who need to work directly with that code.

As with all other aspects of **Extreme Programming (XP)**, the most obvious major win is to make it so close to the code that it is the code. Part of this involves using our good design fundamentals to write clear code and our test suite also plays a key role.

Our TDD tests are code, not manual test documents. They are usually written in the same language and repo as the main code base. They will be written by the same people who are writing the production code – the developers.

The tests are executable. As a form of documentation, you know that something that can run has to be up to date. Otherwise, the compiler will complain, and the code will not run.

Tests also form the perfect example of how to use our production code. They clearly define how it should be set up, what dependencies it has, what its interesting methods and functions are, what its expected effects are, and how it will report errors. Everything you would want to know about that code is in the tests.

It may be surprising at first. Testing and documentation are not normally confused with each other. Because of how TDD works, there is a huge overlap between the two. Our test is a detailed description of what our code should do and how we can make it do that for us.

Summary

In this chapter, we've learned that TDD helps us create good designs, write correct logic, prevent future defects, and provide executable documentation for our code. Understanding what TDD will do for our projects is important to use it effectively and to persuade our teams to use it as well. There are many advantages to TDD and yet it is not used as often as it should be in real-world projects.

In the next chapter, we will look into some common objections to TDD, learn why they are not valid, and how we can help our colleagues overcome them.

Questions and answers

1. What is the connection between testing and clean code?

    There is not a direct one, which is why we need to understand how to write clean code. How TDD adds value is that it forces us to think about how our code will be used before we write it and when it is easiest to clean up. It also allows us to refactor our code, changing its structure without changing its function, with certainty that we have not broken that function.

2. Can tests replace documentation?

    Well-written tests replace some but not all documentation. They become a detailed and up-to-date executable specification for our code. What they cannot replace are documents such as user manuals, operations manuals, or contractual specifications for public **application programming interfaces (APIs)**.

3. What are the problems with writing production code before tests?

    If we write production code first, then add tests later, we are more likely to face the following problems:

    *   Missing broken edge cases on conditionals
    *   Leaking implementation details through interfaces
    *   Forgetting important tests
    *   Having untested execution paths
    *   Creating difficult-to-use code
    *   Forcing more rework when design flaws are revealed later in the process

Further reading

A formal definition of cyclomatic complexity can be found in the WikiPedia link. Basically, every conditional statement adds to the complexity, as it creates a new possible execution path:

```
https://en.wikipedia.org/wiki/Cyclomatic_complexity
```

3
Dispelling Common Myths about TDD

Test-driven development (TDD) brings many benefits to developers and the business. However, it is not always used in real projects. This is something I find surprising. TDD has been demonstrated to improve internal and external code quality in different industrial settings. It works for frontend and backend code. It works across verticals. I have experienced it working in embedded systems, web conferencing products, desktop applications, and microservice fleets.

To better understand how perceptions have gone wrong, let's review the common objections to TDD, then explore how we can overcome them. By understanding the perceived difficulties, we can equip ourselves to be TDD advocates and help our colleagues reframe their thinking. We will examine six popular myths that surround TDD and form constructive responses to them.

In this chapter, we're going to cover the following myths:

- "Writing tests slows me down"
- "Tests cannot prevent every bug"
- "How do you know the tests are right"
- "TDD guarantees good code"
- "Our code is too complex to test"
- "I don't know what to test until I write the code"

Writing tests slows me down

Writing tests slowing development down is a popular complaint about TDD. This criticism has some merit. Personally, I have only ever felt that TDD has made me faster, but academic research disagrees. A meta-analysis of 18 primary studies by the *Association for Computing Machinery* showed that TDD did improve productivity in academic settings but added extra time in industrial contexts. However, that's not the full story.

Understanding the benefits of slowing down

The aforementioned research indicates that the payback for taking extra time with TDD is a reduction in the number of defects that go live in the software. With TDD, these defects are identified and eliminated far sooner than with other approaches. By resolving issues before manual **quality assurance (QA)**, deployment, and release, and before potentially facing a bug report from an end user, TDD allows us to cut out a large chunk of that wasted effort.

We can see the difference in the amount of work to be done in this figure:

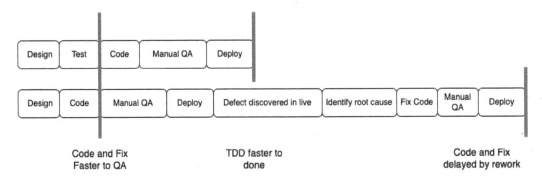

Figure 3.1 – Not using TDD slows us down due to rework

The top row represents developing a feature using TDD, where we have sufficient tests to prevent any defects from going into production. The bottom row represents developing the same feature in a **code-and-fix** style, without TDD, and finding that a defect has gone live in production. Without TDD, we discover faults very late, annoy the user, and pay a heavy time penalty in rework. Note that the code-and-fix solution *looks* like it gets us into the QA stage faster, until we consider all the rework caused by undiscovered defects. The rework is what isn't taken into account in this myth.

Using TDD, we simply make all our design and testing thinking explicit and upfront. We capture and document it using executable tests. Whether we write tests or not, we still spend that same thinking time considering what the specifics that our code needs to cover are. It turns out that the mechanical writing of the test code takes very little time. You can measure that yourself when we write our first test in *Chapter 5, Writing Our First Test*. The total time spent writing a piece of code is the time to design it, plus the time to write the code, plus the time to test it. Even without writing automated tests, the design and coding time remain constant and dominant factors.

The other area conveniently ignored through all this is the time taken to manually test. Without a doubt, our code will be tested. The only question is when and by who. If we write a test first, it is by us, the developers. It happens before any faulty code gets checked into our system. If we leave testing to a manual testing colleague, then we slow down the whole development process. We need to spend time helping our colleague understand what the success criteria are for our code. They must then devise a manual test plan, which often must be written up, reviewed, and accepted into documentation.

Executing manual tests is very time-consuming. Generally, the whole system must be built and deployed to a test environment. Databases must be manually set up to contain known data. The **user interface (UI)** must be clicked through to get to a suitable screen where our new code might be exercised. The output must be manually inspected and a decision made on its correctness. These steps must be manually performed every time we make a change.

Worse still, the later we leave it to test, the greater the chance is that we will have built on top of any faulty code that exists. We cannot know we are doing that, as we haven't tested our code yet. This often becomes difficult to unpick. In some projects, we get so far out of step with the main code branch that developers start emailing patch files to each other. This means we start building on top of this faulty code, making it even harder to remove. These are bad practices but they do occur in real projects.

The contrast to writing a TDD test first could not be greater. With TDD, the setup is automated, the steps are captured and automated, and the result checking is automated. We are talking timescale reductions of minutes for a manual test down to milliseconds using a TDD unit test. This time saving is made every single time we need to run that test.

While manual testing is not as efficient as TDD, there is still one far worse option: no testing at all. Having a defect released to production means that we leave it to our users to test the code. Here, there may be financial considerations and the risk of reputation damage. At the very least, this is a very slow way to discover a fault. Isolating the defective lines of code from production logs and databases is extraordinarily time-consuming. It is also usually frustrating, in my experience.

It's funny how a project that can never find time to write unit tests can *always* find time to trawl production logs, roll back released code, issue marketing communications, and stop all other work to do a **Priority 1 (P1)** fix. Sometimes, it feels like days are easier to find than minutes for some management approaches.

TDD certainly places a time cost up front in writing a test, but in return, we gain fewer faults to rectify in production – with a huge saving in overall cost, time, and reputation compared to multiple rework cycles with defects occurring in live code.

Overcoming objections to tests slowing us down

Build a case that tracks the time spent on undiscovered defects in manual QA and failed deployments. Find some rough figures for the time taken for the most recent live issue to be fixed. Work out which missing unit test could have prevented it. Now work out how long that would have taken to write. Present these figures to stakeholders. It can be even more effective to work out the cost of all that engineering time and any lost revenue.

Knowing that tests do have an overall benefit in terms of fewer defects, let's examine another common objection that tests are of no value, as they cannot prevent every bug.

Tests cannot prevent every bug

A very old objection to testing of any kind is this one: you cannot catch every bug. While this is certainly true, if anything, it means that we need more and better testing, not less. Let's understand the motivations behind this one to prepare an appropriate response.

Understanding why people say tests cannot catch every bug

Straight away, we can agree with this statement. Tests cannot catch every bug. More precisely, it has been proven that testing in software systems can only reveal the presence of defects. It can never prove that no defects exist. We can have many passing tests, and defects can still hide in the places we haven't tested.

This seems to apply in other fields as well. Medical scans will not always reveal problems that are too faint to notice. Wind tunnel tests for aircraft will not always reveal problems under specific flight conditions. Batch sampling in a chocolate factory will not catch every substandard sweet.

Just because we cannot catch every bug, does not mean this invalidates our testing. Every test we write that catches one defect results in one less defect running through our workflow. TDD gives us a process to help us think in terms of testing as we develop, but there are still areas where our tests will not be effective:

- Tests you have not thought to write
- Defects that arise due to system-level interactions

Tests that we have not written are a real problem. Even when writing tests first in TDD, we must be disciplined enough to write a test for every scenario that we want to function. It is easy to write a test and then write the code to make it pass. The temptation is to then just keep adding code because we are on a roll. It is easy to miss an edge case and so not write a test for it. If we have a missing test, we open up the possibility of a defect existing and being found later.

The problem with system-level interactions here refers to the behavior that emerges when you take tested units of software and join them. The interactions between units can sometimes be more complex than anticipated. Basically, if we join up two well-tested things, the new combination itself is still not yet tested. Some interactions have faults that only show up in these interactions, even though the units that they are made up of passed all tests.

These two problems are real and valid. Testing will never cover every possible fault, but this misses the main value of testing. Every test we *do* write will reduce one defect.

By not testing anything, we will never spot anything wrong. We will not prevent any defects. If we test, no matter how little, then we will improve the quality of our code. Every defect that these tests can detect will be prevented. We can see the straw-man nature of this argument: just because we cannot cover every eventuality, it does not mean we should not do what we can.

Overcoming objections to not catching every bug

The way to reframe this is for us to have confidence that TDD prevents many classes of errors from happening. Not all kinds of errors, certainly, but a bank of thousands of tests is going to make a noticeable improvement to the quality of our applications.

To explain this to our colleagues, we can draw on familiar analogies: just because a strong password cannot prevent every hacker, this does not mean we should not use passwords and leave ourselves vulnerable to any and *every* hacker. Staying healthy will not prevent every kind of medical problem but it will prevent many kinds of serious problems.

Ultimately, this is a question of balance. Zero testing is clearly not enough – every single defect will end up going live in this case. We know that testing can never eliminate defects. So, where should we stop? What constitutes *enough*? We can argue that TDD helps us decide on this balance at the best possible time: while we are thinking about writing code. The automated TDD tests we create will save us manual QA time. It's manual work that no longer needs to be done. These time and cost savings compound, repaying us in every single iteration of code.

Now that we understand why testing as much as possible always beats not testing at all, we can look into the next common objection: how do we know the tests themselves were written correctly?

How do you know the tests are right?

This is an objection that has merit, so we need to deeply understand the logic behind it. This is a common objection from people unfamiliar with writing automated tests, as they misunderstand how we avoid incorrect tests. By helping them see the safeguards we put in place, we can help them reframe their thinking.

Understanding the concerns behind writing broken tests

One objection you will hear is, "*How do we know the tests are right if the tests themselves don't have tests?*" This objection was raised the first time I introduced unit tests to a team. It was polarizing. Some of the team understood the value right away. Others were indifferent, but some were actively hostile. They saw this new practice as suggesting they were somehow deficient. It was perceived as a threat. Against that background, one developer pointed out a flaw in the logic I had explained.

I told the team that we could not trust our visual reading of production code. Yes, we are all skilled at reading code, but we are humans, so we miss things. Unit tests would help us avoid missing things. One bright developer asked a great question: if visual inspection does not work for production code, why are we saying that it *does* work for test code? What's the difference between the two?

The right illustration for this came after I needed to test some XML output (which was in 2005, I remember). The code I had written for checking the XML output was truly complex. The criticism was correct. There was no way I could visually inspect that test code and honestly say it was without defects.

So, I applied TDD to the problem. I used TDD to write a utility class that could compare two XML strings and report either that they were the same or what the first difference was. It could be configured to ignore the order of XML elements. I extracted this complex code out of my original test and replaced it with a call to this new utility class. I knew the utility class did not have any defects, as it passed every TDD test that I had written for it. There were many tests, covering every happy path and every edge case I cared about. The original test that had been criticized now became very short and direct.

I asked my colleague who had raised the point to review the code. They agreed that in this new, simpler form, they were happy to agree that the test was correct, visually. They added the caveat "*if the utility class works right.*" Of course, we had the confidence that it passed every TDD test we had written it against. We were certain that it did all the things we specifically wanted it to do, as proven by tests for these things.

Providing reassurance that we test our tests

The essence of this argument is that short, simple code can be visually inspected. To ensure this, we keep most of our unit tests simple and short enough to reason about. Where tests get too complex, we extract that complexity into its own code unit. We develop that using TDD and end up making both the original test code simple enough to inspect and the test utility simple enough for its tests to inspect, a classic example of divide and conquer.

Practically, we invite our colleagues to point out where they feel our test code is too complex to trust. We refactor it to use simple utility classes, these themselves written using simple TDD. This approach helps us build trust, respects the valid concerns of our colleagues, and shows how we can find ways to reduce all TDD tests to simple, reviewable code blocks.

Now that we have addressed knowing our tests are right, another common objection involves having overconfidence in TDD: that simply following the TDD process will therefore guarantee good code. Can that be true? Let's examine the arguments.

TDD guarantees good code

Just as there are often overly pessimistic objections to TDD, here is an opposite view: TDD *guarantees* good code. As TDD is a process, and it claims to improve code, it is quite reasonable to assume that using TDD is all you need to guarantee good code. Unfortunately, that is not at all correct. TDD helps developers write good code and it helps as feedback to show us where we have made mistakes in design and logic. It cannot guarantee good code, however.

Understanding problem-inflated expectations

The issue here is a misunderstanding. TDD is not a set of techniques that directly affect your design decisions. It is a set of techniques that help you specify what you expect a piece of code to do, when, under what conditions, and given a particular design. It leaves you free to choose that design, what you expect it to do, and how you are going to implement that code.

TDD has no suggestions regarding choosing a long variable name over a short one. It does not tell you whether you should choose an interface or an abstract class. Should you choose to split a feature over two classes or five? TDD has no advice there. Should you eliminate duplicated code? Invert a dependency? Connect to a database? Only you can decide. TDD offers no advice. It is not intelligent. *It cannot replace you and your expertise.* It is a simple process, enabling you to validate your assumptions and ideas.

Managing your expectations of TDD

TDD is hugely beneficial in my view but we must regard it in context. It provides instant feedback on our decisions but leaves every important software design decision to us.

Using TDD, we are free to write code using the **SOLID** principles (which will be covered in *Chapter 7, Driving Design — TDD and SOLID*, of this book) or we can use a procedural approach, an object-oriented approach, or a functional approach. TDD allows us to choose our algorithm as we see fit. It enables us to change our minds about how something should be implemented. TDD works across every programming language. It works across every vertical.

Helping our colleagues see past this objection helps them realize that TDD is not some magic system that replaces the intelligence and skill of the programmer. It harnesses this skill by providing instant feedback on our decisions. While this may disappoint colleagues who hoped it would allow perfect code to come from imperfect thinking, we can point out that TDD gives us time to think. The advantage is that it puts thinking and design up front and central. By writing a failing test before writing the production code that makes the test pass, we have ensured that we have thought about what that code should do and how it should be used. That's a great advantage.

Given that we understand that TDD does not design our code for us, yet is still a developer's friend, how can we approach testing complex code?

Our code is too complex to test

Professional developers routinely deal with highly complex code. That's just a fact of life. It leads to one valid objection: our code is too difficult to write unit tests for. The code we work on might be highly valuable, trusted legacy code that brings in significant top-line revenue. This code may be complex. But is it *too* complex to test? Is it true to say that every piece of complex code simply cannot be tested?

Understanding the causes of untestable code

The answer lies in the three ways that code becomes complex and hard to test:

- Accidental complexity: We chose a hard way over a simpler way by accident
- External systems cannot be controlled to set up for our tests
- The code is so entangled that we no longer understand it

Accidental complexity makes code hard to read and hard to test. The best way to think about this is to know that any given problem has many valid solutions. Say we want to add a total of five numbers. We could write a loop. We could create five concurrent tasks that take each number, then report that number to another concurrent task that computes the total (bear with me, please… I've seen this happen). We could have a complex design pattern-based system that has each number trigger an observer, which places each one in a collection, which triggers an observer to add to the total, which triggers an observer every 10 seconds after the last input.

Yes, I know some of those are silly. I just made them up. But let's be honest – what kinds of silly designs have you worked on before? I know I have written code that was more complex than it needed to be.

The key point of the addition of five numbers example is that it really should use a simple loop. Anything else is accidental complexity, neither necessary nor intentional. Why would we do that? There are many reasons. There may be some project constraints, a management directive, or simply a personal preference that steers our decision. However it happened, a simpler solution was possible, yet we did not take it.

Testing more complex solutions generally requires more complex tests. Sometimes, our team thinks it is not worth spending time on that. The code is complex, it will be hard to write tests for, and we think it works already. We think it is best not to touch it.

External systems cause problems in testing. Suppose our code talks to a third-party web service. It is hard to write a repeatable test for that. Our code consumes the external service and the data it sends to us is different each time. We cannot write a test and verify what the service sent us, as we do not know what the service should be sending to us. If we could replace that external service with some dummy service that we could control, then we could fix this problem easily. But if our code does not permit that, then we are stuck.

Entangled code is a further development of this. To write a test, we need to understand what that code does to an input condition: what do we expect the outputs to be? If we have a body of code that we simply do not understand, then we cannot write a test for it.

While these three problems are real, there is one underlying cause to them all: we allowed our software to get into this state. We could have arranged it to only use simple algorithms and data structures. We could have isolated external systems so that we could test the rest of the code without them. We could have modularized our code so that it was not overly entangled.

However, how can we persuade our teams with these ideas?

Reframing the relationship between good design and simple tests

All the preceding problems relate to making software that works yet does not follow good design practices. The most effective way to change this, in my experience, is **pair programming** – working together on the same piece of code and helping each other find these better design ideas. If pair programming is not an option, then code reviews also provide a checkpoint to introduce better designs. Pairing is better as by the time you get to code review, it can be too late to make major changes. It's cheaper, better, and faster to prevent poor design than it is to correct it.

Managing legacy code without tests

We will encounter legacy code without tests that we need to maintain. Often, this code has grown to be quite unmanageable and ideally needs replacing, except that nobody knows what it does anymore. There may be no written documentation or specification to help us understand it. Whatever written material there is may be completely outdated and unhelpful. The original authors of the code may have moved on to a different team or different company.

The best advice here is to simply leave this code alone if possible. Sometimes though, we need to add features that require that code to be changed. Given that we have no existing tests, it is quite likely we will find that adding a new test is all but impossible. The code simply is not split up in a way that gives us access points to hang a test off.

In this case, we can use the **Characterization Test** technique. We can describe this in three steps:

1. Run the legacy code, supplying it with every possible combination of inputs.

2. Record all the outputs that result from each one of these input runs. This output is traditionally called the Golden Master.

3. Write a Characterization Test that runs the code with all inputs again. Compare every output against the captured Golden Master. The test fails if any are different.

This automated test compares any changes that we have made to the code against what the original code did. This will guide us as we refactor the legacy code. We can use standard refactoring techniques combined with TDD. By preserving the defective outputs in the Golden Master, we ensure that we are purely refactoring in this step. We avoid the trap of restructuring the code at the same time as fixing the bugs. When bugs are present in the original code, we work in two distinct phases: first, refactor the code without changing observable behavior. Afterwards, fix the defects as a separate task. We never fix bugs and refactor together. The Characterization Test ensures we do not accidentally conflate the two tasks.

We've seen how TDD helps tackle accidental complexity and the difficulty of changing legacy code. Surely writing a test before production code means we need to know what the code looks like before we test it though? Let's review this common objection next.

I don't know what to test until I write the code

A great frustration for TDD learners is knowing what to test without having written the production code beforehand. This is another criticism that has merit. In this case, once we understand the issue that developers face, we can see that the solution is a technique we can apply to our workflow, not a reframing of thinking.

Understanding the difficulty of starting with testing

To an extent, it's natural to think about how we implement code. It's how we learn, after all. We write `System.out.println("Hello, World!");` instead of thinking up some structure to place around the famous line. Small programs and utilities work just fine when we write them as linear code, similar to a shopping list of instructions.

We begin to face difficulties as programs get larger. We need help organizing the code into understandable chunks. These chunks need to be easy to understand. We want them to be self-documenting and it to be easy for us to know how to call them. The larger the code gets, the less interesting the insides of these chunks are, and the more important the external structure of these chunks – the *outsides* – becomes.

As an example, let's say we are writing a `TextEditorWidget` class, and we want to check the spelling on the fly. We find a library with a `SpellCheck` class in it. We don't care that much about how the `SpellCheck` class works. We only care about how we can *use* this class to check the spelling. We want to know how to create an object of that class, what methods we need to call to get it to do its spellchecking job, and how we can access the output.

This kind of thinking is the definition of software design – how components fit together. It is critical that we emphasize design as code bases grow if we want to maintain them. We use encapsulation to hide the details of data structures and algorithms inside our functions and classes. We provide a simple-to-use programming interface.

Overcoming the need to write production code first

TDD scaffolds design decisions. By writing the test before the production code, we are defining how we want the code under test to be created, called, and used. This helps us see very quickly how well our decisions are working out. If the test shows that creating our object is hard, that shows us that our design should simplify the creation step. The same applies if the object is difficult to use; we should simplify our programming interface as a result.

However, how do we cope with the times when we simply do not yet know what a reasonable design should be? This situation is common when we either use a new library, integrate with some new code from the rest of our team, or tackle a large user story.

To solve this, we use a **spike**, a short section of code that is sufficient to prove the shape of a design. We don't aim for the cleanest code at this stage. We do not cover many edge cases or error conditions. We have the specific and limited goal of exploring a possible arrangement of objects and functions to make a credible design. As soon as we have that, we sketch out some notes on the design and then delete it. Now that we know what a reasonable design looks like, we are better placed to know what tests to write. We can now use normal TDD to drive our design.

Interestingly, when we start over in this way, we often end up driving out a better design than our spike. The feedback loop of TDD helps us spot new approaches and improvements.

We've seen how natural it is to want to start implementing code before tests, and how we can use TDD and spikes to create a better process. We make decisions at the **last responsible moment** – the latest possible time to decide before we are knowingly making an irreversible, inferior decision. When in doubt, we can learn more about the solution space by using a **spike** – a short piece of experimental code designed to learn from and then throw away.

Summary

In this chapter, we've learned six common myths that prevent teams from using TDD and discussed the right approach to reframing those conversations. TDD really deserves a much wider application in modern software development than it has now. It's not that the techniques don't work. TDD simply has an image problem, often among people who haven't experienced its true power.

In the second part of this book, we will start to put the various rhythms and techniques of TDD into practice and build out a small web application. In the next chapter, we will start our TDD journey with the basics of writing a unit test with the **Arrange-Act-Assert (AAA) pattern**.

Questions and answers

1. Why is it believed that TDD slows developers down?

    When we don't write a test, we save the time spent writing the test. What this fails to consider is the extra time costs of finding, reproducing, and fixing a defect in production.

2. Does TDD eliminate human design contributions?

    No. Quite the opposite. We still design our code using every design technique at our disposal. What TDD gives us is a fast feedback loop on whether our design choices have resulted in easy-to-use, correct code.

3. Why doesn't my project team use TDD?

    What a fantastic question to ask them! Seriously. See whether any of their objections have been covered by this chapter. If so, you can gently lead the conversation using the ideas presented.

Further reading

*   https://en.wikipedia.org/wiki/Characterization_test

    More detail on the Characterization Test technique, where we capture the output of an existing software module exactly as-is, with a view to restructuring the code without changing any of its behavior. This is especially valuable in older code where the original requirements have become unclear, or that has evolved over the years to contain defects that other systems now rely on.

*   https://effectivesoftwaredesign.com/2014/03/27/lean-software-development-before-and-after-the-last-responsible-moment/

    An in-depth look at what deciding at the last responsible moment means for software design.

Part 2: TDD Techniques

Part 2 introduces the techniques necessary for effective TDD. Along the way, we will incrementally build the core logic of a word guessing game, Wordz – writing all our tests first.

By the end of this part, we will have produced high-quality code by writing tests first. The SOLID principles and hexagonal architecture will help us organize code into well-engineered building blocks that are easy to test. Test doubles will bring external dependencies under our control. We will look at the bigger picture of test automation and how the test pyramid, QA engineers, and workflow improve our work.

This part has the following chapters:

- *Chapter 4, Building an Application Using TDD*
- *Chapter 5, Writing Our First Test*
- *Chapter 6, Following the Rhythms of TDD*
- *Chapter 7, Driving Design – TDD and SOLID*
- *Chapter 8, Test Doubles – Stubs and Mocks*
- *Chapter 9, Hexagonal Architecture – Decoupling External Systems*
- *Chapter 10, FIRST Tests and the Test Pyramid*
- *Chapter 11, How TDD Fits into Quality Assurance*
- *Chapter 12, Test First, Test Later, Test Never*

4

Building an Application Using TDD

We're going to learn the practical side of TDD by building the application test first. We are also going to use an approach known as **agile software development** as we build. Being agile means building our software in small, self-contained iterations instead of building it all at once. These small steps allow us to learn more about the software design as we go. We adapt and refine the design over time, as we become more certain of how a good design might look. We can offer working functionality to early test users and receive their feedback long before the application is complete. This is valuable. As we have seen in earlier chapters, TDD is an excellent approach for providing rapid feedback on self-contained pieces of software. It is the perfect complement to agile development.

To help us build in this way, this chapter will introduce the technique of **user stories**, which is a way of capturing requirements that fits an agile approach well. We will prepare our Java development environment ready for test-first development before describing what our application will do.

In this chapter, we're going to cover the following topics:

- Introducing the Wordz application
- Exploring agile methods

Technical requirements

The final code for this chapter can be found at `https://github.com/PacktPublishing/Test-Driven-Development-with-Java/tree/main/chapter04`.

To code along – which I highly recommend – we need to set up our development environment first. This will use the excellent JetBrains IntelliJ Java **Integrated Development Environment** (IDE), a free-of-charge Java SDK from Amazon, and some libraries to help us with writing our tests and including the libraries in our Java project. We will assemble all our development tools in the next section.

Preparing our development environment

For this project, we will be using the following tools:

- IntelliJ IDEA IDE 2022.1.3 (Community Edition) or higher

- Amazon Corretto Java 17 JDK

- The JUnit 5 unit test framework

- The AssertJ fluent assertions framework

- The Gradle dependency management system

We will begin by installing our Java IDE, the JetBrains IntelliJ IDE Community Edition, before adding the rest of the tools.

Installing the IntelliJ IDE

To help us work with Java source code, we will use the JetBrains IntelliJ Java IDE, using its free-of-charge Community Edition. This is a popular IDE used in the software industry – and for good reason. It combines an excellent Java editor with auto-completion and code suggestions, together with a debugger, automated refactoring support, Git source control tools, and excellent integration for running tests.

To install IntelliJ, see the following steps:

1. Go to https://www.jetbrains.com/idea/download/.
2. Click on the tab for your operating system.
3. Scroll down to the **Community** section.
4. Follow the installation instructions for your operating system.

Once complete, the IntelliJ IDE should be installed on your computer. The next step is to create an empty Java project, using the Gradle package management system, and then set up whichever version of Java we wish to use. The installations for Mac, Windows, and Linux are usually straightforward.

Setting up the Java project and libraries

Once IntelliJ is installed, we can import the starter project provided in the accompanying GitHub repository. This will set up a Java project that uses the Amazon Corretto 17 **Java Development Kit (JDK)**, the JUnit 5 unit test runner, the Gradle build management system, and the AssertJ fluent assertions library.

To do this, see the following steps:

1. In your web browser, go to `https://github.com/PacktPublishing/Test-Driven-Development-with-Java`.

2. Use your preferred `git` tool to clone the whole repository on your computer. If you use the `git` command-line tool, this will be as follows:

    ```
    git clone https://github.com/PacktPublishing/Test-Driven-Development-with-Java.git
    ```

3. Launch IntelliJ. You should see the welcome screen:

Figure 4.1 – IntelliJ welcome screen

4. Click **Open** and then navigate to the `chapter04` folder of the repository that we just cloned. Click to highlight it:

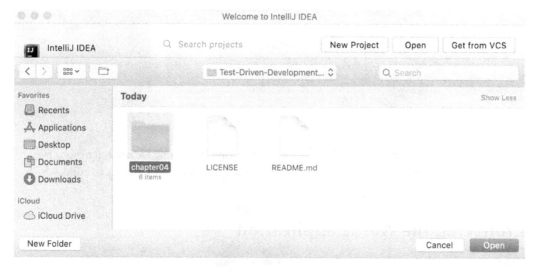

Figure 4.2 – Select the code folder

5. Click the **Open** button.

6.   Wait for IntelliJ to import the files. You should see this workspace open:

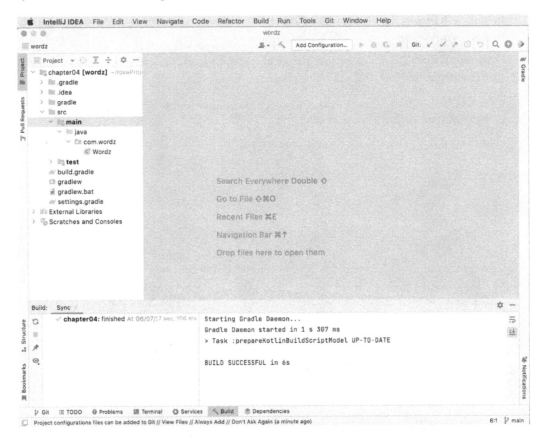

Figure 4.3 – IntelliJ workspace view

We now have the IDE set up with a skeleton project containing everything we need to make a start. In the next section, we will describe the main features of the application we are going to build, which we will start to do in the next chapter.

Introducing the Wordz application

In this section, we will describe the application that we are going to build at a high level, before going on to look at the agile process we will use to build it. The application is called Wordz and it is based on a popular word guessing game. Players try to guess a five-letter word. Points are scored based on how quickly a player guesses the word. The player gets feedback on each guess to steer them towards the right answer. We are going to build the server-side components of this application throughout the remainder of this book using various TDD techniques.

Describing the rules of Wordz

To play Wordz, a player will have up to six attempts to guess a five-letter word. After each attempt, letters in the word are highlighted as follows:

- The correct letter in the correct position has a black background

- The correct letter in the wrong position has a gray background

- Incorrect letters not present in the word have a white background

The player can use this feedback to make a better next guess. Once a player guesses the word correctly, they score some points. They get six points for a correct guess on the first attempt, five points for a correct guess on the second attempt, and one point for a correct guess on the sixth and final attempt. Players compete against each other in various rounds to gain the highest score. Wordz is a fun game as well as a gentle brain workout.

Whilst building a user interface is outside the scope of this book, it is very helpful to see a possible example:

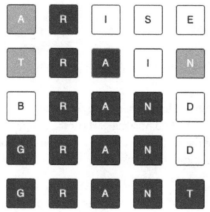

Figure 4.4 – The Wordz game

Technically, we are going to create the backend web service component for this game. It will expose an **Application Programming Interface** (**API**) so that a user interface can use the service and will keep track of the game state in a database.

To focus on the techniques of TDD, we will leave certain things out of our scope, such as user authentication and the user interface. A production version would, of course, include these aspects. But to implement these features, we don't need any new TDD techniques.

This simple design will allow us to fully explore TDD through all the layers of a typical web application.

Now that we've defined what we're going to build, the next section will introduce the development approach we will use to build it.

Exploring agile methods

As we build Wordz, we are going to use an iterative approach, where we build the application as a series of features that our users can work with. This is known as **agile development**. It is effective as it allows us to ship features to users earlier and on a regular schedule. It allows us as developers to learn more about the problems we are solving and how a good software design looks as we go. This section will compare the benefits of agile development to waterfall approaches, then introduce an agile requirements gathering tool called user stories.

The predecessor to agile is called **waterfall development**. It is called this because the project stages flow as a waterfall does, each one is fully completed before the next one is begun.

In a waterfall project, we split development into sequential stages:

1. Collecting requirements
2. Performing an analysis of requirements
3. Creating a complete software design
4. Writing all the code
5. Testing the code

In theory, every stage is perfectly executed, everything works, and there are no problems. In reality, there are always problems.

We discover certain requirements we had missed. We find that the design documents cannot be coded exactly as they were written. We find missing parts of the design. The coding itself can run into difficulties. The worst part is that the end user never sees any working software until the very end. If what they see is not what they had in mind, we have a very expensive set of changes and reworking to do.

The reason for this is that *humans have limited foresight*. Try as we might, we cannot predict the future with any accuracy. I can sit here with a hot cup of coffee and know accurately that it will go cold in twenty minutes. But I can't tell you what the weather will be three months from now. Our ability to predict the future is limited to short time frames, for processes with clear-cut causes and effects.

Waterfall development performs very poorly in the face of uncertainty and change. It is designed around the notion that all things can be known and planned in advance. A better approach is to *embrace* change and uncertainty, making it an active part of the development process. This is the basis of agile development. At its core lies an iterative approach, where we take one small feature that our users care about, then build that feature completely, allowing our users to try it out. If changes are needed, we do another iteration of development. The costs of change are much lower when our development process actively supports change.

Professional agile development processes rely on maintaining one single code base that is always tested and represents *the best version to date* of our software. This code is always ready to deploy to users. We grow this code base one feature at a time, continuously improving its design as we go.

Techniques such as TDD play a major role in this, by ensuring our code is well designed and thoroughly tested. Every time we commit code to the main trunk, we already know it has passed many TDD tests. We know we are happy with its design.

To better support iterative development, we choose an iterative technique for capturing requirements. This technique is called user stories, which we will describe in the next section.

Reading user stories – the building block of planning

As development is iterative and embraces refactoring and reworking, it makes sense that the old methods of specifying requirements won't work. We are no longer served by thousands of pages of requirements set in stone up front. We are better served by taking one requirement at a time, building it, and learning from it. Over time, we can prioritize the features users want and learn more about how a *good design* will look.

Through agile techniques, we do not have to know the future in advance; we can discover it alongside our users.

Supporting this change is a new way to express requirements. Waterfall projects start with a complete requirements document, detailing every feature formally. The complete set of requirements – often thousands of them – were expressed in formal language such as "*The system shall…*" and then the details were explained in terms of changes to the software system. With agile development, we don't want to capture requirements in that way. We want to capture them following two key principles:

- Requirements are presented one at a time in isolation
- We emphasize the value to the user, not the technical impact on the system

The technique for doing this is called the user story. The first user story to tackle for Wordz looks as follows:

AS A PLAYER
I WANT TO SCORE MY ATTEMPT
SO THAT I CAN IMPROVE MY NEXT GUESS

Figure 4.5 – The user story

The format of a user story is always the same – it comprises three sections:

- As a [person or machine that uses the software], …
- I want [a specific outcome from that software] …
- … so that [a task that is important is achieved].

The three sections are written this way to emphasize that agile development centers around value delivered to the users of the system. These are not technical requirements. They do not (indeed, *must not*) specify a solution. They simply state which user of the system should get what valuable outcome out of it.

The first part always starts with "*As a ….*" It then names the user role that this story will improve. This can be any user – whether human or machine – of the system. The one thing it must never be is the system itself, as in, "*As a system.*" This is to enforce clear thinking in our user stories; they must always deliver some benefit to some user of the system. They are never an end in themselves.

To give an example from a photo-taking app, as developers, we might want a technical activity to optimize photo storage. We might write a story such as, "*As a system, I want to compact my image data to optimize storage.*" Instead of writing from a technical viewpoint, we can reframe this to highlight the benefit to the user: "*As a photographer, I want fast access to my stored photographs and to maximize space for new ones.*"

The "*I want…*" section describes the desired outcome the user wants. It is always described in user terminology, not technical terminology. Again, this helps us focus on what our users want our software to achieve for them. It is the purest form of capturing the requirements. There is no attempt made at this stage to suggest how anything will be implemented. We simply capture what it is that the user sets out to do.

The final part, "*…so that…*", provides context. The "*As a …*" section describes *who* benefits, the "*I want…*" section describes *how* they benefit, and the "*…so that…*" section describes *why* they need this feature. This forms the justification for the time and costs required for developing this feature. It can be used to prioritize which features to develop next.

This user story is where we start development. The heart of the Wordz application is its ability to evaluate and score the player's current guess at a word. It's worth looking at how this work will proceed.

Combining agile development with TDD

TDD is a perfect complement to agile development. As we learned in earlier chapters, TDD helps us improve our design and prove that our logic is correct. Everything we do is aimed at delivering working software to our users, without defects, as quickly as possible. TDD is a great way to achieve this.

The workflow we will use is typical for an agile TDD project:

1. Pick a user story prioritized for impact.
2. Think a little about the design to aim for.
3. Use TDD to write the application logic in the core.
4. Use TDD to write code to connect the core to a database.
5. Use TDD to write code to connect to an API endpoint.

This process repeats. It forms the rhythm of writing the core application logic under a unit test, then growing the application outward, connecting it to API endpoints, user interfaces, databases, and external web services. Working this way, we retain a lot of flexibility within our code. We can also work quickly, concentrating upfront on the most important parts of our application code.

Summary

We've learned the key ideas that let us build an application iteratively, getting value at each step and avoiding a *big design up front* approach that often disappoints. We can read user stories, which will drive building our TDD application in small, well-defined steps. We now also know the process we will use to build our application – using TDD to get a thoroughly tested, central core of clean code, and then drive out connections to the real world.

In the next chapter, we'll make a start on our application. We will learn the three key components of every TDD test by writing our first test and making sure it passes.

Questions and answers

1. Waterfall development sounds as though it should work well – why doesn't it?

   Waterfall development would work well if we knew about every missing requirement, every change request from the users, every bad design decision, and every coding error at the start of the project. But humans have limited foresight, and it is impossible to know these things in advance. So, waterfall projects never work smoothly. Expensive changes crop up at a later stage of the project – just when you don't have the time to address them.

2. Can we do agile development without TDD?

   Yes, although that way, we miss out on the advantages of TDD that we've covered in previous chapters. We also make our job harder. An important part of Agile development is always demonstrating the latest working code. Without TDD, we need to add a large manual test cycle into our process. This slows us down significantly.

Further reading

- *Mastering React Test-Driven Development, ISBN 9781789133417*

   If you would like to build a user interface for the Wordz application, using the popular React web UI framework is an excellent way to do it. This Packt book is one of my personal favorites. It shows how to apply the same kind of TDD techniques we are using server-side into frontend work. It also explains React development from the ground up in a highly readable way.

- *Agile Model-Based Systems Engineering Cookbook, ISBN 9781838985837*

   This book provides further details on how to craft effective user stories and other useful techniques for capturing agile requirements, modeling, and analysis.

5
Writing Our First Test

It's time for us to dive in and write our first TDD unit test in this chapter. To help us do this, we will learn about a simple template that helps us organize each test into a logical, readable piece of code. Along the way, we will learn some key principles we can use to make our tests effective. We will see how writing the test first forces us to make decisions about the design of our code and its ease of use, before needing to think about implementation details.

After some examples covering those techniques, we will make a start on our Wordz application, writing a test first before adding production code to make that test pass. We will use the popular Java unit testing libraries JUnit5 and AssertJ to help us write easy-to-read tests.

In this chapter, we will cover the following main principles behind writing effective unit tests:

- Starting TDD: **Arrange-Act-Assert**
- Defining a good test
- Catching common errors
- Asserting exceptions
- Only testing public methods
- Learning from our tests
- Beginning Wordz – our first test

Technical requirements

The final code in this chapter can be found at `https://github.com/PacktPublishing/Test-Driven-Development-with-Java/tree/main/chapter05`.

Starting TDD: Arrange-Act-Assert

Unit tests are nothing mysterious. They're just code, executable code written in the same language that you write your application in. Each unit test forms the first use of the code you want to write. It calls the code just as it will be called in the real application. The test executes that code, captures all the outputs that we care about, and checks that they are what we expected them to be. Because the test uses our code in the exact same way that the real application will, we get instant feedback on how easy or difficult our code is to use. This might sound obvious, and it is, but it is a powerful tool to help us write clean and correct code. Let's take a look at an example of a unit test and learn how to define its structure.

Defining the test structure

It's always helpful to have templates to follow when we do things and unit tests are no exception. Based on commercial work done on the Chrysler Comprehensive Compensation Project, TDD inventor Kent Beck found that unit tests had certain features in common. This became summarized as a recommended structure for test code, called **Arrange-Act-Assert** or **AAA**.

> **The original definition of AAA**
>
> The original description of AAA can be found here, in the C2 wiki: `http://wiki.c2.com/?ArrangeActAssert`.

To explain what each section does, let's walk through a completed unit test for a piece of code where we want to ensure that a username is displayed in lowercase:

```java
import org.junit.jupiter.api.Test;
import static org.assertj.core.api.Assertions.*;

public class UsernameTest {

    @Test
    public void convertsToLowerCase() {
        var username = new Username("SirJakington35179");

        String actual = username.asLowerCase();

        assertThat(actual).isEqualTo("sirjakington35179");
    }

}
```

The first thing to notice is the class name for our test: `UsernameTest`. This is the first piece of storytelling for readers of our code. We are describing the behavioral area we are testing, in this case, usernames. All our tests, and indeed all our code, should follow this storytelling approach: what do we want the readers of our code to understand? We want them to clearly see what the problem that we are solving is and how the code that solves it should be used. We want to demonstrate to them that the code works correctly.

The unit test itself is the `convertsToLowerCase()` method. Again, the name describes what we expect to happen. When the code runs successfully, the username will be converted to lowercase. The names are intentionally simple, clear, and descriptive. This method has the `@Test` annotation from the **JUnit5** test framework. The annotation tells JUnit that this is a test that it can run for us.

Inside the `@Test` method, we can see our *Arrange-Act-Assert* structure. We first *arrange* for our code to be able to run. This involves creating any objects required, supplying any configuration needed, and connecting any dependent objects and functions. Sometimes, we do not need this step, for example, if we are testing a simple standalone function. In our example code, the *Arrange* step is the line that creates the `username` object and supplies a name to the constructor. It then stores that object ready to use in the local `username` variable. It is the first line of the `var username = new Username("SirJakington35179");` test method body.

The *Act* step follows. This is the part where we cause our code under test to act – we run that code. This is always a call to the code under test, supplying any necessary parameters, and arranging to capture the results. In the example, the `String actual = username.asLowerCase();` line is the *Act* step. We call the `asLowerCase()` method on our `username` object. It takes no parameters and returns a simple `String` object containing the lowercase text `sirjakington35179` as a result.

Completing our test is the final *Assert* step. The `assertThat(actual).isEqualTo("sirjakington35179");` line is our *Assert* step here. It uses the `assertThat()` method and the `isEqualTo()` method from the `AssertJ` fluent assertions library. Its job is to check whether the result we returned from the *Act* step matches our expectations or not. Here, we are testing whether all the uppercase letters in the original name have been converted to lowercase.

Unit tests like this are easy to write, easy to read, and they run very quickly. Many such tests can run in under 1 second.

The `JUnit` library is the industry-standard unit test framework for Java. It provides us with a means to annotate Java methods as unit tests, lets us run all our tests, and visually displays the results, as shown here in the *IntelliJ* IDE window:

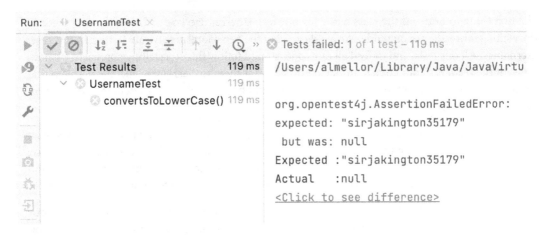

Figure 5.1 – Output from the JUnit test runner

We see here that the unit test failed. The test expected the result to be the `sirjakington35179` text string but instead, we received `null`. Using TDD, we would complete just enough code to make that test pass:

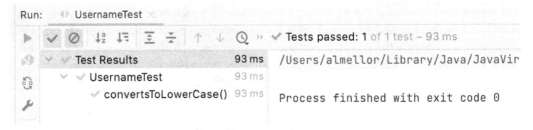

Figure 5.2 – A JUnit test pass

We can see that our change to the production code has made this test pass. It has *gone green*, to use the popular term. Tests that fail are described as red tests and those that pass are green. This is based on the colors shown in popular IDEs, which are based on traffic signals in turn. Seeing all these short iterations of red tests turning to green is surprisingly satisfying, as well as building confidence in our work. The tests help us focus on the design of our code by forcing us to work backward from outcomes. Let's look at what this means.

Working backward from outcomes

One thing we notice right away is just how unimportant the actual code that makes this test pass is. Everything in this test is about defining the expectations of that code. We are setting boundaries around *why* our code is useful and *what* we expect it to do. We are not constraining *how* it does it in any way. We are taking an *outside-in* view of code. Any implementation that makes our test pass is acceptable.

This seems to be a transition point in learning to use TDD. Many of us learned to program by writing implementations first. We thought about how the code would work. We went deep into the algorithms and data structures behind a specific implementation. Then, as a last thought, we wrapped it all up in some kind of callable interface.

TDD turns this on its head. We intentionally design our callable interface first, as this is what the users of that code will see. We use the test to precisely describe how the code will be set up, how it will be called, and what we can expect it to do for us. Once we get used to doing this outside-in design first, TDD follows very naturally and improves our workflow efficiency in several important ways. Let's review what these improvements are.

Increasing workflow efficiency

Unit tests like these increase our efficiency as developers in several ways. The most obvious is that the code we write has passed a test: we know it works. We are not waiting around for a manual QA process to find a defect and then raise a bug report for rework in the future. We find and fix bugs *now*, before ever releasing them into the main source trunk, let alone to users. We have documented our intentions for our colleagues. If anyone wants to know how our Username class works, it is right there in the test – how you create the object, which methods you can call, and what we expect the outcomes to be.

Unit tests give us a way to run code in isolation. We are no longer forced to rebuild a whole application, run it, set up test data entries in our database, log in to the user interface, navigate to the correct screen, and then visually inspect the output of our code. We run the test. That's it. This allows us to execute code that is not yet fully integrated into our application's main trunk. This speeds up our work. We can get started more quickly, spend more time on developing the code at hand, and spend less time on cumbersome manual testing and deployment processes.

A further benefit is that this act of design improves the modularity of our code. By designing code that can be tested in small pieces, we remind ourselves to write code that can execute in small pieces. That has been the basic approach to design since the 1960s and remains as effective today as it ever was.

This section has covered the standard structure that we use to organize every unit test but it doesn't guarantee that we will write a good test. To achieve this, each test needs to have particular properties. The **FIRST** principles describe the properties of a good test. Let's learn how to apply these next.

Defining a good test

Like all code, unit test code can be written in better or worse ways. We've seen how AAA helps us structure a test correctly and how accurate, descriptive names tell the story of what we intend our code to do. The most useful tests also follow the FIRST principles and use one assert per test.

Applying the FIRST principles

These are a set of five principles that make tests more effective:

- Fast
- Isolated
- Repeatable
- Self-verifying
- Timely

Unit tests need to be **fast**, just as our earlier example was. This is especially important for test-first TDD, as we want that immediate feedback while we explore our design and implementation. If we run a unit test, and it takes even as little as 15 seconds to complete, we will soon stop running tests as often. We will degenerate into writing big chunks of production code without tests so that we spend less time waiting for slow tests to finish. This is the exact opposite of what we want from TDD, so we work hard to keep tests fast. We need unit tests to run in 2 seconds or less, ideally milliseconds. Even two seconds is really quite a high number.

Tests need to be **isolated** from one another. This means that we can pick any test or any combination of tests and run them in any order we like and always get the same result. One test must not depend on another test having been run before it. This is often a symptom of failing to write fast tests, so we compensate by caching results or arranging step setups. This is a mistake, as it slows down development, especially for our colleagues. The reason is that we don't know the special order in which the tests must run. When we run any test on its own, and if it has not been properly isolated, it will fail as a false negative. That test no longer tells us anything about our code under test. It only tells us that we have not run some other test before it, without telling us which test that might be. Isolation is critical to a healthy TDD workflow.

Repeatable tests are vital to TDD. Whenever we run a test with the same production code, that test must always return the same pass or fail result. This might sound obvious but care needs to be taken to achieve this. Think about a test that checks a function that returns a random number between 1 and 10. If we assert that the number seven is returned, this test will only pass occasionally, even if we have correctly coded the function. In this regard, three popular sources of misery are tests involving the database, tests against time, and tests through the user interface. We will explore techniques to handle these situations in *Chapter 8, Test Doubles –Stubs and Mocks*.

All tests must be **self-verifying**. This means we need executable code to run and check whether the outputs are as expected. *This step must be automated*. We must not leave this check to manual inspection, perhaps by writing the output to a console and having a human check it against a test plan. Unit tests derive huge value from being automated. The computer checks the production code, freeing us from the tedium of following a test plan, the slowness of human activities, and the likelihood of human error.

Timely tests are tests written at just the right time to be most useful. The ideal time to write a test is just before writing the code that makes that test pass. It's not unusual to see teams use less beneficial approaches. The worst one, of course, is to never write any unit tests and rely on manual QA to find bugs. With this approach, we get none of the design feedback available. The other extreme is to have an analyst write every test for the component – or even the whole system – upfront, leaving the coding as a mechanical exercise. This also fails to learn from design feedback. It can also result in overspecified tests that *lock in* poor design and implementation choices. Many teams start by writing some code and then go on to write a unit test, thereby missing out on an opportunity for early design feedback. It can also lead to untested code and faulty edge case handling.

We've seen how the FIRST principles help us focus on crafting a good test. Another important principle is not to try to test too much all at once. If we do, the test becomes very difficult to understand. A simple solution to this is to write a single assert per test, which we will cover next.

Using one assert per test

Tests provide the most useful feedback when they are short and specific. They act as a microscope working on the code, each test highlighting one small aspect of our code. The best way to ensure this happens is by writing one assertion per test. This prevents us from tackling too much in one test. This focuses on the error messages we get during test failures and helps us control the complexity of our code. It forces us to break things down a little further.

Deciding on the scope of a unit test

Another common misunderstanding is what a *unit* means in a unit test. The unit refers to the test isolation itself – each test can be considered a standalone unit. As a result, the size of the code under test can vary a lot, as long as that test can run in isolation.

Thinking of the test itself as the unit unifies several popular opinions about what the scope of a unit test should be. Often, it is said that the unit is the smallest piece of testable code – a function, method, class, or package. All of these are valid options. Another common argument is that a unit test should be a class test – one unit test class per production code class, with one unit test method per production method. While common, this isn't usually the best approach. It unnecessarily couples the structure of the test to the structure of the implementation, making the code *harder* to change in the future, not easier.

The ideal goal of a unit test is to cover one *externally visible behavior*. This applies at several different scales in the code base. We can unit test an entire user story across multiple packages of classes, provided we can avoid manipulating external systems such as a database or the user interface. We'll look into techniques for doing that in *Chapter 9, Hexagonal Architecture – Decoupling External Systems*. We often also use unit tests that are *closer* to the details of the code, testing only the public methods of a single class.

Once we have written our test based on the design that we would like our code to have, we can concentrate on the more obvious aspect of testing: verifying that our code is correct.

Catching common errors

The traditional view of testing is of it as a process to check that code works as it is intended to work. Unit tests excel at this and automate the process of running the code with known inputs and checking for expected outputs. As we are human, all of us make mistakes from time to time as we write code and some of these can have significant impacts. There are several common simple mistakes we can make and unit tests excel at catching them all. The most likely errors are the following:

- Off-by-one errors
- Inverted conditional logic
- Missing conditions
- Uninitialized data
- The wrong algorithm
- Broken equality checks

As an example, going back to our earlier test for a lowercase username, suppose we decided not to implement this using the `String` built-in `.toLowerCase()` method, but instead tried to roll our own loop code, like this:

```
public class Username {
    private final String name;

    public Username(String username) {
        name = username;
    }

    public String asLowerCase() {
        var result = new StringBuilder();

        for (int i=1; i < name.length(); i++) {
            char current = name.charAt(i);
            if (current > 'A' && current < 'Z') {
                result.append(current + 'a' - 'A');
            } else {
                result.append( current );
```

```
        }
      }

      return result.toString() ;
   }
}
```

We would see right away that this code isn't correct. The test fails, as shown in the following figure:

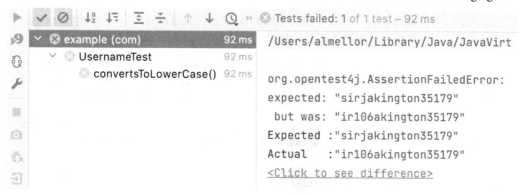

```
Tests failed: 1 of 1 test – 92 ms

example (com)              92 ms      /Users/almellor/Library/Java/JavaVirt
  UsernameTest             92 ms
    convertsToLowerCase()  92 ms      org.opentest4j.AssertionFailedError:
                                      expected: "sirjakington35179"
                                       but was: "ir106akington35179"
                                      Expected :"sirjakington35179"
                                      Actual   :"ir106akington35179"
                                      <Click to see difference>
```

Figure 5.3 – A common coding error

The first error in this code is a simple off-by-one error – the first letter is missing from the output. That points to an error in initializing our loop index but there are other errors in this code as well. This test reveals two defects. Further tests would reveal two more. Can you see what they are by visual inspection alone? How much more time and effort is it to analyze code like this in our heads, rather than using automated tests?

Asserting exceptions

One area where unit tests excel is in testing error handling code. As an example of testing *exception throwing*, let's add a business requirement that our usernames must be at least four characters long. We think about the design we want and decide to throw a custom exception if the name is too short. We decide to represent this custom exception as class InvalidNameException. Here's what the test looks like, using AssertJ:

```
@Test
public void rejectsShortName() {
    assertThatExceptionOfType(InvalidNameException.class)
        .isThrownBy(()->new Username("Abc"));
}
```

We can consider adding another test specifically aimed at proving that a name of four characters is accepted and no exception is thrown:

```
@Test
public void acceptsMinimumLengthName() {
    assertThatNoException()
            .isThrownBy(()->new Username("Abcd"));
}
```

Alternatively, we may simply decide that this explicit test is not needed. We may cover it implicitly with other tests. It is a good practice to add both tests to make our intentions clear.

The test names are fairly general, starting with either `rejects` or `accepts`. They describe the outcome that the code is being tested for. This allows us to change our minds about the error handling mechanics later, perhaps switching to something other than exceptions to signal the error.

Unit tests can catch common programming errors and verify error handling logic. Let's look at a major principle of writing our unit tests to give us maximum flexibility when implementing our methods.

Only testing public methods

TDD is all about testing the behaviors of components, not their implementations. As we have seen in our test in the previous section, having a test for the behavior we want enables us to choose any implementation that will do the job. We focus on what's important – *what* a component does – not on the less important details – *how* it does it.

Inside a test, this appears as calling public methods or functions on public classes and packages. The public methods are the behaviors we choose to expose to the wider application. Any private data or supporting code in classes, methods, or functions remain hidden.

A common mistake that developers make when learning TDD is that they make things public just to simplify testing. Resist the temptation. A typical mistake here is to take a private data field and expose it for testing using a public getter method. This weakens the encapsulation of that class. It is now more likely that the getter will be misused. Future developers may add methods to other classes that really belong in this one. The design of our production code is important. Fortunately, there is a simple way of preserving encapsulation without compromising testing.

Preserving encapsulation

If we feel we need to add getters to all our private data so that the test can check that each one is as expected, it is often better to treat this as a **value object**. A value object is an object that lacks identity. Any two value objects that contain the same data are considered to be equal. Using value objects, we can make another object containing the same private data and then test that the two objects are equal.

In Java, this requires us to code a custom `equals()` method for our class. If we do this, we should also code a `hashcode()` method, as the two go hand in hand. Any implementation that works will do. I recommend using the `Apache commons3` library, which uses Java reflection capabilities to do this:

```
@Override
public boolean equals(Object other) {
    return EqualsBuilder.reflectionEquals(this, other);
}

@Override
public int hashCode() {
    return HashCodeBuilder.reflectionHashCode(this);
}
```

You can find out more about these library methods at `https://commons.apache.org/proper/commons-lang/`.

Simply adding those two methods (and the `Apache commons3` library) to our class means that we can keep all our data fields private and still check that all the fields have the expected data in them. We simply create a new object with all the expected fields, then assert that it is equal to the object we are testing.

As we write each test, we are using the code under test for the first time. This allows us to learn a lot about how easy our code is to use, allowing us to make changes if we need to.

Learning from our tests

Our tests are a rich source of feedback on our design. As we make decisions, we write them as test code. Seeing this code – the first usage of our production code – brings into sharp focus how good our proposed design is. When our design isn't good, the AAA sections of our test will reveal those design issues as code smells in the test. Let's try to understand in detail how each of these can help identify a faulty design.

A messy Arrange step

If the code in our Arrange step is messy, our object may be difficult to create and configure. It may need too many parameters in a constructor or too many optional parameters left as `null` in the test. It may be that the object needs too many dependencies injected, indicating that it has too many responsibilities or it might need too many primitive data parameters to pass in a lot of configuration items. These are signals that the way we create our object might benefit from a redesign.

A messy Act step

Calling the main part of the code in the Act step is usually straightforward but it can reveal some basic design errors. For example, we might have unclear parameters that we pass in, signatures such as a list of `Boolean` or `String` objects. It is very hard to know what each one means. We could redesign this by wrapping those difficult parameters in an easy-to-understand new class, called a **configuration object**. Another possible problem is if the Act step requires multiple calls to be made in a specific order. That is error-prone. It is easy to call them in the wrong order or forget one of the calls. We could redesign to use a single method that wraps all of this detail.

A messy Assert step

The Assert step will reveal whether the results of our code are difficult to use. Problem areas might include having to call accessors in a specific order or perhaps returning some conventional code smells, such as an array of results where every index has a different meaning. We can redesign to use safer constructs in either case.

In each of these cases, one of the sections of code in our unit test looks wrong – it has a code smell. That is because the design of the code we are testing has the same code smell. This is what is meant by unit tests giving fast feedback on design. They are the first user of the code we are writing, so we can identify problem areas early on.

We now have all the techniques we need to start writing our first test for our example application. Let's make a start.

Limitations of unit tests

One very important idea is that an automated test can only prove the presence of a defect, not the absence. What this means is that if we think of a boundary condition, write a test for that, and the test fails, we know we have a defect in our logic. However, if all our tests pass, that *does not and cannot* mean our code is free of defects. It only means that our code is free of all the defects that we have thought to test for. There simply is no magic solution that can ensure our code is defect-free. TDD gives us a significant boost in that direction but we must never claim our code is defect-free just because all our tests pass. This is simply untrue.

One important consequence of this is that our QA engineering colleagues remain as important as they ever were, although we now help them start from an easier standing point. We can deliver TDD-tested code to our manual QA colleagues, and they can be assured that many defects have been prevented and proven to be absent. This means that they can start work on manual exploratory testing, finding all the things we never thought to test. Working together, we can use their defect reports to write further unit tests to rectify what they find. The contribution of QA engineers remains vital, even with TDD. We need all the help our team can get in our efforts to write high-quality software.

Code coverage – an often-meaningless metric

Code coverage is a measure of how many lines of code have been executed in a given run. It is measured by instrumenting the code and this is something that a code coverage tool will do for us. It is often used in conjunction with unit testing to measure how many lines of code were executed while running the test suite.

In theory, you can see how this might mean that missing tests can be discovered in a scientific way. If we see that a line of code was not run, we must have a missing test somewhere. That is both true and helpful but the converse is not true. Suppose we get 100% code coverage during our test run. Does that mean the software is now completely tested and correct? No.

Consider having a single test for an `if (x < 2)` statement. We can write a test that will cause this line to execute and be included in code coverage reports. However, a single test is not enough to cover all the possibilities of behavior. The conditional statement might have the wrong operator – less than instead of less than or equal to. It might have the incorrect limit of 2 when it should be 20. Any single test cannot fully explore the combinations of behavior in that statement. We can have code coverage tell us that the line has been run and that our single test passed but we can still have several logic errors remaining. We can have 100% code coverage and still have missing tests.

Writing the wrong tests

Time for a short personal story about how my best attempt at TDD went spectacularly wrong. In a mobile application that calculated personal tax reports, there was a particular yes/no checkbox in the app to indicate whether you had a student loan or not, since this affects the tax you pay. It had six consequences in our application and I thoroughly TDD tested each one, carefully writing my tests first.

Sadly, I had misread the user story. I had inverted every single test. Where the checkbox should apply the relevant tax, it now did not apply it, and vice versa.

This was thankfully picked up by our QA engineer. Her only comment was that she could find absolutely no workaround in the system for this defect. We concluded that TDD had done an excellent job of making the code do what I wanted it to do but I had done a rather less excellent job of figuring out what that should be. At least it was a very quick fix and retest.

Beginning Wordz

Let's apply these ideas to our Wordz application. We're going to start with a class that will contain the core of our application logic, one that represents a word to guess and that can work out the score for a guess.

We begin by creating a unit test class and this immediately puts us into software design mode: what should we call the test? We'll go with `WordTest`, as that outlines the area we want to cover – the word to be guessed.

Typical Java project structures are divided into packages. The production code lives under `src/main/java` and the test code is located under `src/test/java`. This structure describes how production and test code are equally important parts of the source code, while giving us a way to compile and deploy only the production code. We always ship test code with the production code when we are dealing with source code, but for deployed executables, we only omit the tests. We will also follow the basic Java package convention of having a unique name for our company or project at the top level. This helps avoid clashes with library code. We'll call ours `com.wordz`, named after the application.

The next design step is to decide which behavior to drive out and test first. We always want a simple version of a happy path, something that will help drive out the normal logic that will most commonly execute. We can cover edge cases and error conditions later. To begin with, let's write a test that will return the score for a single letter that is incorrect:

1. Write the following code to begin our test:

    ```
    public class WordTest {

        @Test
        public void oneIncorrectLetter() {


        }
    }
    ```

    The name of the test gives us an overview of what the test is doing.

2. To start our design, we decide to use a class called `Word` to represent our word. We also decide to supply the word to guess as a constructor parameter to our object instance of class `Word` we want to create. We code these design decisions into the test:

    ```
    @Test
    public void oneIncorrectLetter () {
        new Word("A");
    }
    ```

3. We use autocomplete at this point to create a new `Word` class in its own file. Double-check in `src/main folder tree` and not `src/test`:

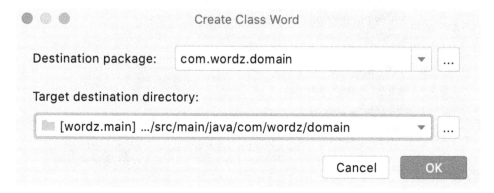

Figure 5.4 – Creating a class dialog

4. Click **OK** to create the file in the source tree inside the right package.

5. Now, we rename the `Word` constructor parameter:

```java
public class Word {
    public Word(String correctWord) {
// No Action
    }
}
```

6. Next, we return to the test. We capture the new object as a local variable so that we can test it:

```java
@Test
public void oneIncorrectLetter () {
    var word = new Word("A");
}
```

The next design step is to think of a way to pass a guess into the `Word` class and return a score.

7. Passing the guess in is an easy decision – we'll use a method that we'll call `guess()`. We can code these decisions into the test:

```java
@Test
public void oneIncorrectLetter () {
    var word = new Word("A");

    word.guess("Z");
}
```

8. Use autocomplete to add the `guess()` method to the `Word` class:

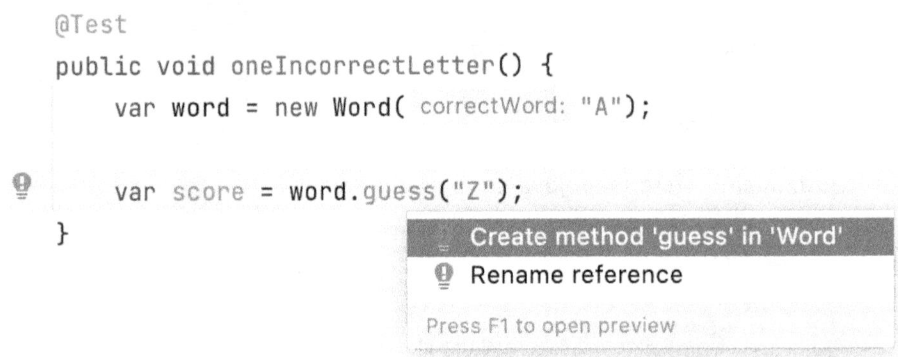

    ```
    @Test
    public void oneIncorrectLetter() {
        var word = new Word( correctWord: "A");

        var score = word.guess("Z");
    }
    ```

    | Create method 'guess' in 'Word' |
    | Rename reference |
    | Press F1 to open preview |

Figure 5.5 – Creating the Word class

9. Click *Enter* to add the method, then change the parameter name to a descriptive name:

    ```
    public void guess(String attempt) {

    }
    ```

10. Next, let's add a way to get the resulting score from that guess. Start with the test:

    ```
    @Test
    public void oneIncorrectLetter () {
        var word = new Word("A");

        var score = word.guess("Z");
    }
    ```

    Then, we need a little think about what to return from the production code.

We probably want an object of some sort. This object must represent the score from that guess. Because our current user story is about the scores for a five-letter word and the details of each letter, we must return one of *exactly right*, *right letter*, *wrong place*, or *letter not present*.

There are several ways to do this and now is the time to stop and think about them. Here are some viable approaches:

- A class with five getters, each one returning an `enum`.
- A **Java 17** `record` type with the same getters.
- A class with an `iterator` method, which iterates over five `enum` constants.

- A class with an `iterator` method that returns one interface for each letter score. The scoring code would implement a concrete class for each type of score. This would be a purely object-oriented way of adding a *callback* for each possible outcome.

- A class that iterated over results for each letter and you passed in a `Java 8 lambda` function for each of the outcomes. The correct one would be called as a callback for each letter.

That's already a lot of design options. The key part of TDD is that we are considering this *now* before we write any production code. To help us decide, let's sketch out what the calling code will look like. We need to consider plausible extensions to the code – will we need more or fewer than five letters in a word? Would the scoring rules ever change? Should we care about any of those things *right now*? Would the people reading this code in the future more easily grasp any one of these ideas than the others? TDD gives us fast feedback on our design decisions and that forces us to take a design workout right now.

One overriding decision is that we will not return the colors that each letter should have. That will be a UI code decision. For this core domain logic, we will return only the fact that the letter is *correct*, in the *wrong position*, or *not present*.

It's easy enough with TDD to sketch out the calling code because it *is* the test code itself. After about 15 minutes of pondering what to do, here are the three design decisions we will use in this code:

- Supporting a variable number of letters in a word

- Representing the score using a simple enum of INCORRECT, PART_CORRECT, or CORRECT

- Accessing each score by its position in the word, zero-based

These decisions support the **KISS** principle, usually termed **keep it simple, stupid**. The decision to support a variable number of letters does make me wonder whether I've overstepped another principle – **YAGNI** – or **you ain't gonna need it**. In this case, I'm convincing myself that it's not too much of a speculative design and that the readability of the `score` object will make up for that. Let's move on to the design:

1. Capture these decisions in the test:

```
@Test
public void oneIncorrectLetter() {
    var word = new Word("A");

    var score = word.guess("Z");

    var result = score.letter(0);
    assertThat(result).isEqualTo(Letter.INCORRECT);
}
```

We can see how this test has locked in those design decisions about how we will use our objects. It says *nothing at all* about how we will implement those methods internally. *This is critical to effective TDD.* We have also captured and documented all the design decisions in this test. Creating an **executable specification** such as this is an important benefit of TDD.

2.   Now, run this test. Watch it fail. This is a surprisingly important step.

     We might think at first that we only ever want to see passing tests. This is not totally true. Part of the work in TDD is having confidence that your tests are working. Seeing a test fail when we know we have not written the code to make it pass yet gives us confidence that our test is probably checking the right things.

3.   Let's make that test pass, by adding code to `class Word`:

```java
public class Word {
    public Word(String correctWord) {
        // Not Implemented
    }

    public Score guess(String attempt) {
        var score = new Score();
        return score;
    }
}
```

4.   Next, create `class Score`:

```java
public class Score {
    public Letter letter(int position) {
        return Letter.INCORRECT;
    }
}
```

Again, we used IDE shortcuts to do most of the work in writing that code for us. The test passes:

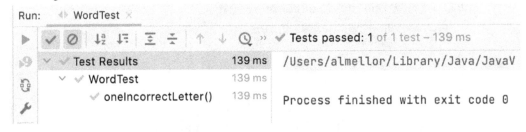

Figure 5.6 – A test passing in IntelliJ

We can see that the test passed and that it took 0.139 seconds to run. That certainly beats any manual test.

We also have a repeatable test, which we can run for the remainder of the project life cycle. The time saving compared to manual testing will add up every time we run the test suite.

You will notice that although the test passes, the code seems like it is cheating. The test only ever expects `Letter.INCORRECT` and the code is hardcoded to always return that. It clearly could never possibly work for any other values! This is expected at this stage. Our first test has set out a rough design for the interface of our code. It has not yet begun to drive out the full implementation. We will do that with our subsequent tests. This process is called *triangulation*, where we rely on adding tests to drive out the missing implementation details. By doing this, all our code is covered by tests. We get 100% *meaningful* code coverage for free. More importantly, it breaks our work down into smaller chunks, creates progress with frequent deliverables, and can lead to some interesting solutions.

Another thing to notice is that our one test led us to create two classes, covered by that one test. This is highly recommended. Remember that our unit test covers a behavior, not any specific implementation of that behavior.

Summary

We've taken our first steps into TDD and learned about the AAA structure of each test. We've seen how it is possible to design our software and write our test before our production code and get cleaner, more modular designs as a result. We learned what makes for a good test and learned some common techniques used to catch common programming errors and test code that throws exceptions.

It is important to understand the flow of using AAA sections inside our FIRST tests, as this gives us a template we can reliably follow. It is also important to understand the flow of design ideas, as used in the previous Wordz example. Writing our tests is literally taking the design decisions we make and capturing them in unit test code. This provides fast feedback on how clean our design is, as well as providing an executable specification for future readers of our code.

In the next chapter, we will add tests and drive out a complete implementation for our word-scoring object. We will see how TDD has a rhythm that drives work forward. We will use the **Red, Green, Refactor** approach to keep refining our code and keep both code and tests clean without overengineering them.

Questions and answers

1. How do we know what test to write if we have no code to test?

   We reframe this thinking. Tests help us design a small section of code upfront. We decide what interface we want for this code and then capture these decisions in the AAA steps of a unit test. We write just enough code to make the test compile, and then just enough to make the test run and fail. At this point, we have an executable specification for our code to guide us as we go on to write the production code.

2. Must we stick to one test class per production class?

No, and this is a common misunderstanding when using unit tests. The goal of each test is to specify and run a behavior. This behavior will be implemented in some way using code – functions, classes, objects, library calls, and the like – but this test in no way constrains how the behavior is implemented. Some unit tests test only one function. Some have one test per public method per class. Others, like in our worked example, give rise to more than one class to satisfy the test.

3. Do we always use the AAA structure?

It's a useful recommendation to start out that way but we sometimes find that we can omit or collapse a step and improve the readability of a test. We might omit the Arrange step, if we had nothing to create for, say, a static method. We may collapse the Act step into the Assert step for a simple method call to make the test more readable. We can factor our common Arrange step code into a *JUnit* `@BeforeEach` annotate method.

4. Are tests throwaway code?

No. They are treated with the same importance and care as production code. The test code is kept clean just as the production code is kept clean. The readability of our test code is paramount. We must be able to skim-read a test and quickly see why it exists and what it does. The test code is not deployed in production but that does not make it any less important.

Following the Rhythms of TDD

6

We've seen how individual unit tests help us explore and capture design decisions about our code and keep our code defect-free and simple to use, but that's not all they can do. TDD has rhythms that help us with the whole development cycle. By following the rhythms, we have a guide on what to do next at each step. It is helpful to have this technical structure that allows us to think deeply about engineering good code and then capture the results.

The first rhythm was covered in the last chapter. Inside each test, we have a rhythm of writing the Arrange, Act, and Assert sections. We'll add some detailed observations on succeeding with this next. We'll go on to cover the larger rhythm that guides us as we refine our code, known as the **red, green, refactor (RGR)** cycle. Together, they help us craft our code to be easy to integrate into the broader application and made of clean, simple-to-understand code. Applying these two rhythms ensures that we deliver high-quality code at pace. It provides us with several small milestones to hit during each coding session. This is highly motivating, as we gain a sense of steady progress toward our goal of building our application.

In this chapter, we're going to cover the following topics:

- Following the RGR cycle
- Writing our next tests for Wordz

Technical requirements

The final code in this chapter can be found at `https://github.com/PacktPublishing/Test-Driven-Development-with-Java/tree/main/chapter06`. It is recommended to follow along with the exercise by typing the code in yourself – and thinking about all the decisions we will be making as we go.

Following the RGR cycle

We saw in the previous chapter how a single unit test is split into three parts, known as the Arrange, Act, and Assert sections. This forms a simple rhythm of work that guides us through writing every test. It forces us to design how our code is going to be used – the outside of our code. If we think of an object as being an encapsulation boundary, it makes sense to talk about what is inside and outside that boundary. The public methods form the outside of our object. The Arrange, Act and Assert rhythm helps us design those.

We're using the word *rhythm* here in an almost musical sense. It's a constant, repeating theme that holds our work together. There is a regular flow of work in writing tests, writing code, improving that code, and then deciding which test to write next. Every test and piece of code will be different, but the rhythm of work is the same, as though it were a steady beat in an ever-changing song.

Once we have written our test, we turn to creating the code that is inside our object – the private fields and methods. For this, we make use of another rhythm called RGR. This is a three-step process that helps us to build confidence in our test, create a basic implementation of our code, and then refine it safely.

In this section, we will learn what work needs to be done in each of the three phases.

Starting on red

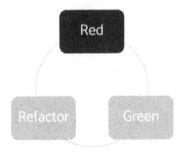

Figure 6.1 – The red phase

We always start with the first phase called the red phase. The goal of this phase is to use the Arrange, Act and Assert template to get our test up and running and ready to test the code we will write next. The most important part of this phase is to make sure that the test does not pass. We call this a failing test, or a red test, due to the color that most graphical test tools use to indicate a failing test.

That's rather counter-intuitive, isn't it? We normally aim to make things work right the first time in development. However, we want our test to fail at this stage to give us confidence that it is working correctly. If the test passes at this point, it's a concern. Why does it pass? We know that we have not yet written any of the code we are testing. If the test passes now, that means we either do not need to write any new code or we have made a mistake in the test. The *Further reading* section has a link to eight reasons why a test might not be running correctly.

The most common mistake here is getting the assertion wrong. Identify the error and fix it before moving on. We must have that red test in place so that we can see it change from failing to passing as we correctly add code.

Keep it simple – moving to green

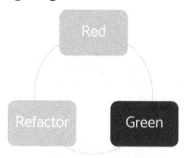

Figure 6.2 – The green phase

Once we have our failing test, we are free to write the code that will make it pass. We call this the production code – the code that will form part of our production system. We treat our production code as a *black-box* component. Think of an integrated circuit in electronics, or perhaps some kind of mechanical sealed unit. The component has an inside and an outside. The inside is where we write our production code. It is where we hide the data and algorithms of our implementation. We can do this using any approach we choose – object-oriented, functional, declarative, or procedural. Anything we fancy. The outside is the **Application Programming Interface** (**API**). This is the part we use to connect to our component and use it to build bigger pieces of software. If we choose an object-oriented approach, this API will be made of public methods on an object. With TDD, the first piece we connect to is our test, and that gives us fast feedback on how easy the connection is to use.

The following diagram shows the different pieces – the inside, outside, test code, and other users of our component:

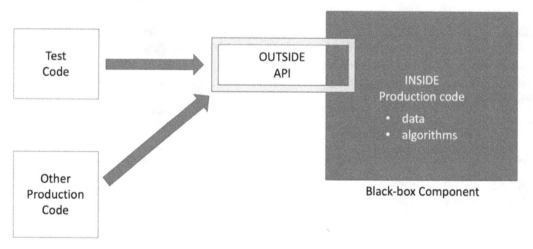

Figure 6.3 – The inside and outside a black-box component

Because our implementation is encapsulated, we can change our minds about it later as we learn more without breaking the test.

There are two guidelines for this phase:

- **Use the simplest code that could possibly work**: Using the simplest code is important. There can be a temptation to use over-engineered algorithms, or perhaps use the latest language feature just for an excuse to use it. Resist this temptation. At this stage, our goal is to get the test to pass and nothing more.

- **Don't overthink the implementation details**: We don't need to overthink this. We don't need to write the perfect code on our first attempt. We can write a single line, a method, several methods, or entirely new classes. We will improve this code in the next step. Just remember to make the test pass and not go beyond what this test is covering in terms of functionality.

Refactoring to clean code

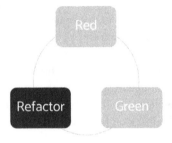

Figure 6.4 – The refactor phase

This is the phase where we go into software engineering mode. We have some working, simple code with a test that passes. Now is the time to refine that into clean code – meaning code that will be easy to read later. With the confidence that a passing test provides, we are free to apply any valid refactoring technique to our code. Some examples of refactoring techniques we can use during this phase include the following:

- Extracting a method to remove duplicated code

- Renaming a method to express what it does better

- Renaming a variable to express what it contains better

- Splitting a long method into several smaller ones

- Extracting a smaller class

- Combining a long parameter list into its own class

All these techniques have one goal: to make our code easier to understand. This will make it easier to maintain. Remember to keep that green test passing throughout these changes. By the end of this phase, we will have a unit test covering a piece of production code that we have engineered to be easy to work with in the future. That's a good place to be.

Now we're familiar with what to do in each phase of the RGR cycle, let's apply that to our Wordz application.

Writing our next tests for Wordz

So, what should we write for our next tests? What would be a useful and small-enough step so that we do not fall into the trap of writing beyond what our tests can support? In this section, we will continue building out the Wordz application scoring system using TDD. We will discuss how we choose to move forward at each step.

For the next test, a good choice is to play it safe and move only a small step further. We will add a test for a single correct letter. This will drive out our first piece of genuine application logic:

1. Let's start on red. Write a failing test for a single, correct letter:

```
@Test
public void oneCorrectLetter() {
    var word = new Word("A");

    var score = word.guess("A");

    assertThat(score.letter(0))
        .isEqualTo(Letter.CORRECT);
}
```

This test is intentionally similar to the one before. The difference is that it tests for a letter being correct, rather than being incorrect. We have used the same word – a single letter, "A" – intentionally. This is important when writing tests – use test data that helps to tell the story of what we are testing and why. The story here is that the same word with a different guess will lead to a different score – obviously key to the problem we are solving. Our two test cases completely cover both possible outcomes of any guess of a single-letter word.

Using our IDE auto-completion features, we quickly arrive at changes to class Word.

2. Now let's move to green by adding the production code to make the test pass:

```
public class Word {
    private final String word;

    public Word(String correctWord) {
```

```
        this.word = correctWord;
    }


    public Score guess(String attempt) {
        var score = new Score(word);


        score.assess( 0, attempt );
        return score;
    }

}
```

The goal here is to get the new test to pass while keeping the existing test passing. We don't want to break any existing code. We've added a field called word, which will store the word we are supposed to be guessing. We've added a public constructor to initialize this field. We have added code into the guess() method to create a new Score object. We decide to add a method to this Score class called assess(). This method has the responsibility of assessing what our guess should score. We decide that assess() should have two parameters. The first parameter is a zero-based index for which letter of the word we wish to assess a score. The second parameter is our guess at what the word might be.

We use the IDE to help us write class Score:

```
public class Score {
    private final String correct;
    private Letter result = Letter.INCORRECT ;


    public Score(String correct) {
        this.correct = correct;
    }


    public Letter letter(int position) {
        return result;
    }


    public void assess(int position, String attempt) {
        if ( correct.charAt(position) == attempt.
            charAt(position)){
            result = Letter.CORRECT;
        }
```

```
        }
    }
```

To cover the new behavior tested by the `oneCorrectLetter()` test, we add the preceding code. Instead of the `assess()` method always returning `Letter.INCORRECT` as it did previously, the new test has forced a new direction. The `assess()` method must now be able to return the correct score when a guessed letter is correct.

To achieve this, we added a field called `result` to hold the latest score, code to return that result from the `letter()` method, and code into the `assess()` method to check whether the first letter of our guess matches the first letter of our word. If we have got this right, both of our tests should now pass.

Run all the tests to see how we are doing:

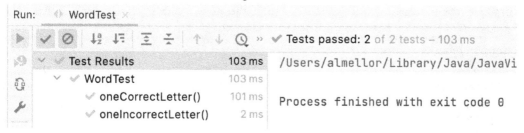

Figure 6.5 – Two tests passing

There's a lot to review here. Notice how both of our tests are passing. By running all the tests so far, we have proven that we have not broken anything. The changes we made to our code added the new feature and did not break any existing features. That's powerful. Take note of another obvious aspect – we know our code works. We do not have to wait until a manual test phase, wait until some integration point, or wait until the user interface is ready. We know our code works now. As a minor point, note the time duration of 0.103 seconds. The two tests were completed in one-tenth of one second, much faster than testing this manually. Not bad at all.

Design-wise, we have moved on. We have moved past the hard-coded `Letter.INCORRECT` result with code that can detect both correct and incorrect guesses. We have added the important design concept of an `assess()` method into `class Score`. This is significant. Our code now reveals a design; the `Score` object will know the correct `word` and will be able to use the `assess()` method against the guess, `attempt`. The terminology used here forms a good description of the problem we are solving. We want to assess a guess to return a word score.

Now that the test passes, we can move on – but an important part of TDD is continuously improving our code and working toward a better design, guided by tests. We now enter the refactor phase of the RGR cycle. Once again, TDD hands control back to us. Do we want to refactor? What things should we refactor? Why? Is it worth doing this right now or can we defer this until a later step?

Let's review the code and look for **code smells**. A code smell is an indication that the implementation may need improving. The name comes from the idea of the smell that food has once it starts to go off.

One code smell is duplicated code. Alone, a little duplicated code might be okay. But it is an early warning that perhaps too much copy-and-paste has been used, and that we have failed to capture an important concept more directly. Let's review our code to eliminate duplication. We can also look for two other common code smells – unclear naming, and blocks of code that would be easier to read if they were extracted out into their own method. Obviously, this is subjective, and we will all have different views on what to change.

Defining code smells

The term code smell originally appeared on the C2 wiki. It's worth a read to see the given examples of code smells. It has a helpful definition that notes a code smell is something that needs review but may not necessarily need to be changed:

`https://wiki.c2.com/?CodeSmell`.

Let's reflect on the inside of the `assess()` method. It just seems cluttered with too much code. Let's extract a helper method to add some clarity. We can always revert the change if we feel it doesn't help.

3. Let's refactor. Extract an `isCorrectLetter()` method for clarity:

```
public void assess(int position, String attempt) {
    if (isCorrectLetter(position, attempt)){
        result = Letter.CORRECT;
    }
}


private boolean isCorrectLetter(int position,
                                String attempt) {
    return correct.charAt(position) ==
        attempt.charAt(position);
}
```

Once more, we run all the tests to prove this refactoring has not broken anything. The tests pass. In the preceding code, we have split out a complex conditional statement into its own private method. The motivation was to get a method name into the code. This is an effective way of commenting on our code – in a way that the compiler helps us keep up to date. It helps the calling code in the `assess()` method tell a better story. The `if` statement now says "if this is a correct letter" more or less in English. That is a powerful aid to readability.

> **Readability happens during writing not reading**
>
> A common question from coding beginners is "How can I improve my ability to read code?"
>
> This is a valid question, as any line of code will be read by human programmers many more times than it was written. Readability is won or lost when you write the code. Any line of code can be written either to be easy to read or hard to read. We get to choose as writers. If we consistently choose ease of reading over anything else, others will find our code easy to read.
>
> Badly written code is hard to read. Sadly, it is easy to write.

There are two more areas I want to refactor at this stage. The first is a simple method to improve test readability.

Let's refactor the test code to improve its clarity. We will add a *custom* `assert` method:

```java
@Test
public void oneCorrectLetter() {
    var word = new Word("A");

    var score = word.guess("A");

    assertScoreForLetter(score, 0, Letter.CORRECT);
}

private void assertScoreForLetter(Score score,
                  int position, Letter expected) {
    assertThat(score.letter(position))
        .isEqualTo(expected);
}
```

The preceding code has taken the `assertThat()` assertion and moved it into its own private method. We have called this method `assertScoreForLetter()` and given it a signature that describes what information is needed. This change provides a more direct description of what the test is doing while reducing some duplicated code. It also protects us against changes in the implementation of the assertion. This seems to be a step toward a more comprehensive assertion, which we will need once we support guesses with more letters. Once again, instead of adding a comment to the source code, we have used a method name to capture the intent of the `assertThat()` code. Writing *AssertJ custom matchers* are another way of doing this.

The next refactoring we may want to do is a little more controversial, as it is a design change. Let's do the refactoring, discuss it, then possibly revert the code if we don't like it. That will save hours of wondering about what the change would look like.

4. Let's change how we specify the letter position to check in the `assess()` method:

```java
public class Score {
    private final String correct;
    private Letter result = Letter.INCORRECT ;
    private int position;

    public Score(String correct) {
        this.correct = correct;
    }

    public Letter letter(int position) {
        return result;
    }

    public void assess(String attempt) {
        if (isCorrectLetter(attempt)){
            result = Letter.CORRECT;
        }
    }

    private boolean isCorrectLetter(String attempt) {
        return correct.charAt(position) == attempt.
        charAt(position);
    }
}
```

We've removed the `position` parameter from the `assess()` method and converted it into a field called `position`. The intention is to simplify the usage of the `assess()` method. It no longer needs to explicitly state which position is being assessed. That makes the code easier to call. The code we have just added will only work in the case where the position is zero. This is fine, as this is the only thing required by our tests at this stage. We will make this code work for non-zero values later.

The reason this is a controversial change is that it requires us to change the test code to reflect that change in the method signature. I am prepared to accept this, knowing that I can use my IDE-automated refactoring support to do this safely. It also introduces a risk: we must ensure that position is set to the correct value before we call `isCorrectLetter()`. We'll see how this develops. This may make the code more difficult to understand, in which case the simplified `assess()` method probably will not be worth it. We can change our approach if we find this to be the case.

We are now at a point where the code is complete for any single-letter word. What should we attempt next? It seems as though we should move on to two-letter words and see how that changes our tests and logic.

Advancing the design with two-letter combinations

We can proceed to add tests aimed at getting the code to handle two-letter combinations. This is an obvious step to take after getting the code to work with a single letter. To do this, we will need to introduce a new concept into the code: a letter can be present in the word, but not in the position we guessed it to be:

1. Let's begin by writing a test for a second letter that is in the wrong position:

    ```
    @Test
    void secondLetterWrongPosition() {
        var word = new Word("AR");
        var score = word.guess("ZA");
        assertScoreForLetter(score, 1,
                        Letter.PART_CORRECT);
    }
    ```

    Let's change the code inside the assess() method to make this pass and keep the existing tests passing.

2. Let's add initial code to check all the letters in our guess:

    ```
    public void assess(String attempt) {
        for (char current: attempt.toCharArray()) {
            if (isCorrectLetter(current)) {
                result = Letter.CORRECT;
            }
        }
    }


    private boolean isCorrectLetter(char currentLetter) {
        return correct.charAt(position) == currentLetter;
    }
    ```

The main change here is to assess all of the letters in attempt and not assume it only has one letter in it. That, of course, was the purpose of this test – to drive out this behavior. By choosing to convert the attempt string into an array of char, the code seems to read quite well. This simple algorithm iterates over each char, using the current variable to represent the current letter to be assessed. This requires the `isCorrectLetter()` method to be refactored for it to accept and work with the `char` input – well, either that or converting `char` to a `String`, and that looks ugly.

The original tests for single-letter behaviors still pass, as they must. We know the logic inside our loop cannot possibly be correct – we are simply overwriting the `result` field, which can only store a result for one letter at most. We need to improve that logic, but we won't do that until we have added a test for that. Working this way is known as **triangulation** – we make the code more general-purpose as we add more specific tests. For our next step, we will add code to detect when our attempted letter occurs in the word in some other position.

3.   Let's add code to detect when a correct letter is in the wrong position:

```java
public void assess(String attempt) {
    for (char current: attempt.toCharArray()) {
        if (isCorrectLetter(current)) {
            result = Letter.CORRECT;
        } else if (occursInWord(current)) {
            result = Letter.PART_CORRECT;
        }
    }
}

private boolean occursInWord(char current) {
    return
        correct.contains(String.valueOf(current));
}
```

We've added a call to a new private method, `occursInWord()`, which will return `true` if the current letter occurs anywhere in the word. We have already established that this current letter is not in the right place. This should give us a clear result for a correct letter not in the correct position.

This code makes all three tests pass. Immediately, this is suspicious, as it shouldn't happen. We already know that our logic overwrites the single `result` field and this means that many combinations will fail. What has happened is that our latest test is fairly weak. We could go back and strengthen that test, by adding an extra assertion. Alternatively, we can leave it as it is and write another test. Dilemmas such as this are common in development and it's not usually worth spending too much time thinking about them. Either way will move us forward.

Let's add another test to completely exercise the behavior around the second letter being in the wrong position.

4. Add a new test exercising all three scoring possibilities:

```
@Test
void allScoreCombinations() {
    var word = new Word("ARI");
    var score = word.guess("ZAI");
    assertScoreForLetter(score, 0, Letter.INCORRECT);
    assertScoreForLetter(score, 1,
                            Letter.PART_CORRECT);
    assertScoreForLetter(score, 2, Letter.CORRECT);
}
```

As expected, this test fails. The reason is obvious upon inspecting the production code. It's because we were storing results in the same single-valued field. Now that we have a failing test for that, we can correct the scoring logic.

5. Add a `List` of results to store the result for each letter position separately:

```
public class Score {
    private final String correct;
    private final List<Letter> results =
                            new ArrayList<>();
    private int position;

    public Score(String correct) {
        this.correct = correct;
    }

    public Letter letter(int position) {
        return results.get(position);
    }

    public void assess(String attempt) {
        for (char current: attempt.toCharArray()) {
            if (isCorrectLetter(current)) {
                results.add(Letter.CORRECT);
            } else if (occursInWord(current)) {
                results.add(Letter.PART_CORRECT);
            } else {
```

```
                    results.add(Letter.INCORRECT);
        }


        position++;
    }
}


private boolean occursInWord(char current) {
    return
      correct.contains(String.valueOf(current));
}


private boolean isCorrectLetter(char
   currentLetter) {
    return correct.charAt(position) ==
            currentLetter;
    }
}
```

This took a couple of attempts to get right, driven by failures in the test we just added. The preceding end result passes all four tests, proving it can correctly score all combinations in a three-letter word. The main change was to replace the single-valued result field with an ArrayList of results and change the letter(position) implementation method to use this new collection of results. Running that change caused a failure, as the code could no longer detect an incorrect letter. Previously, that had been handled by the default value of the result field. Now, we must do that explicitly for each letter. We then need to update the position within the loop to track which letter position we are assessing.

We've added a test, watched it go red and fail, then added code to make the test go green and pass, so now it is time to refactor. There are things about both the test and the production code that don't seem quite right.

In the production code class Score, it is the loop body of the assess() method that seems unwieldy. It has a long loop body with logic in it and a set of if-else-if blocks. It feels as though the code could be made clearer. We can extract the loop body into a method. The method name then gives us a place to describe what is happening to each thing. The loop then becomes shorter and simpler to grasp. We can also replace the if-else-if ladders with a simpler construct.

6. Let's extract the logic inside the loop body into a `scoreFor()` method:

```
public void assess(String attempt) {
    for (char current: attempt.toCharArray()) {
        results.add( scoreFor(current) );
        position++;
    }
}


private Letter scoreFor(char current) {
    if (isCorrectLetter(current)) {
        return Letter.CORRECT;
    }

    if (occursInWord(current)) {
        return Letter.PART_CORRECT;
    }

    return Letter.INCORRECT;
}
```

This reads far more clearly. The body of the `scoreFor()` method is now a concise description of the rules for scoring each letter. We replaced the `if-else-if` construction with a simpler `if-return` construction. We work out what the score is, then exit the method immediately.

The next job is to clean up the test code. In TDD, test code is given equal priority to production code. It forms part of the documentation about the system. It needs to be maintained and extended alongside the production code. We treat test code readability with the same importance as production code.

The code smell with the test code is around the asserts. Two things could be improved. There is an obvious duplication in the code that we could eliminate. There is also a question about how many assertions should be made in one test.

7. Let's remove the duplicated assertion code by extracting a method:

```
@Test
void allScoreCombinations() {
    var word = new Word("ARI");
    var score = word.guess("ZAI");
    assertScoreForGuess(score, INCORRECT,
```

```
                                        PART_CORRECT,
                                        CORRECT);
}


private void assertScoreForGuess(Score score, Letter...
    for (int position=0;
            position < expectedScores.length;
            position++){
        Letter expected = expectedScores[position];

        assertThat(score.letter(position))
            .isEqualTo(expected);
    }
}
```

By extracting the assertScoreForGuess() method, we create a way to check the scores for a variable number of letters. This eliminates those copy-pasted assert lines that we had and raises the level of abstraction. The test code reads more clearly as we now describe tests in terms of the order of INCORRECT, PART_CORRECT, CORRECT that we expect the score to be in. By adding a static import to those enums, syntax clutter is also beneficially reduced.

The earlier tests can now be manually modified to make use of this new assertion helper. This allows us to inline the original assertScoreForLetter() method, as it no longer adds value.

8.   Now, let's take a look at the final set of tests following our refactoring:

```
package com.wordz.domain;


import org.junit.jupiter.api.Test;
import static com.wordz.domain.Letter.*;
import static org.assertj.core.api.Assertions.assertThat;


public class WordTest {
    @Test
    public void oneIncorrectLetter() {
        var word = new Word("A");
        var score = word.guess("Z");
        assertScoreForGuess(score, INCORRECT);
    }


    @Test
```

```
    public void oneCorrectLetter() {
        var word = new Word("A");
        var score = word.guess("A");
        assertScoreForGuess(score, CORRECT);
    }

    @Test
    public void secondLetterWrongPosition() {
        var word = new Word("AR");
        var score = word.guess("ZA");
        assertScoreForGuess(score,   INCORRECT,
                                     PART_CORRECT);
    }

    @Test
    public void allScoreCombinations() {
        var word = new Word("ARI");
        var score = word.guess("ZAI");
        assertScoreForGuess(score,   INCORRECT,
                                     PART_CORRECT,
                                     CORRECT);
    }


    private void assertScoreForGuess(Score score,
        Letter... expectedScores) {
        for (int position = 0;
                position < expectedScores.length;
                position++) {
            Letter expected =
                    expectedScores[position];
            assertThat(score.letter(position))
                    .isEqualTo(expected);
        }
    }
}
```

This appears to be a comprehensive set of test cases. Every line of production code has been driven out as a direct result of adding a new test to explore a new aspect of behavior. The test code seems easy to read and the production code also seems clearly implemented and simple to call. The test forms an executable specification of the rules for scoring a guess at a word.

That's achieved everything we set out to at the start of this coding session. We have grown the capability of our `Score` class using TDD. We have followed the RGR cycle to keep both our test code and production code following good engineering practices. We have robust code, validated by unit tests, and a design that makes this code easy to call from our wider application.

Summary

In this chapter, we have applied the RGR cycle to our code. We've seen how this splits the work into separate tasks, which results in confidence in our test, a rapid path to simple production code, and less time spent to improve the maintainability of our code. We've looked at removing code smells from both the production code and the test code. As part of our work in this chapter, we've used ideas that help us move ahead and decide what tests we should write next. The techniques in this chapter enable us to write multiple tests and incrementally drive out the detailed logic in our production code.

In the next chapter, we're going to learn about some object-oriented design ideas known as the SOLID principles, enabling us to use TDD to grow our application still further.

Questions and answers

1. What are the two key rhythms of TDD?

    Arrange, Act, Assert, and RGR. The first rhythm helps us write the body of the test while designing the interface to our production code. The second rhythm works to help us create and then refine the implementation of that production code.

2. How can we write tests before code?

    Instead of thinking about how we are going to implement some code, we think about how we are going to call that code. We capture those design decisions inside a unit test.

3. Should tests be throwaway code?

    No. In TDD, unit tests are given equal weight to the production code. They are written with the same care and are stored in the same code repository. The only difference is that the test code itself will not be present in the delivered executable.

4. Do we need to refactor after every test pass?

    No. Use this time as an opportunity to decide what refactoring is needed. This applies to both the production code and the test code. Sometimes, none is needed and we move on. Other times, we sense that a larger change would be beneficial. We might choose to defer that larger change until later once we have more code in place.

Further reading

- *Getting Green on Red*

  An article by Jeff Langr describing eight different ways a test can pass for the wrong reasons. If we're aware of these issues, we can avoid them as we work.

  ```
  https://medium.com/pragmatic-programmers/3-5-getting-green-on-
  red-d189240b1c87
  ```

- *Refactoring: Improving the design of existing code*, Martin Fowler (ISBN 978-0134757599)

  The definitive guide to refactoring code. The book describes step-by-step transformations of code that preserve its behavior but improve clarity. Interestingly, most transformations come in pairs, such as the pair of techniques known as *Extract Method* and *Inline Method*. This reflects the trade-offs involved.

- AssertJ documentation for custom matchers

  This chapter briefly mentioned *AssertJ custom matchers*. These are very useful ways of creating reusable customized assertions for your code. These assertion classes are themselves unit-testable and can be written using test-first TDD. For that reason alone, they are superior to adding a private method to handle a customized assertion.

  The following link provides many examples provided by the AssertJ distribution on github.

  ```
  https://github.com/assertj/assertj-examples/tree/main/assertions-
  examples/src/test/java/org/assertj/examples/custom
  ```

7
Driving Design – TDD and SOLID

So far, we've created some basic unit tests that have driven out a simple design for a couple of classes. We've experienced how **test-driven development** (**TDD**) makes decision-making about design choices central. In order to build out to a larger application, we are going to need to be able to handle designs of greater complexity. To do this, we are going to apply some recommended approaches to assessing what makes one design preferable to another.

The SOLID principles are five design guidelines that steer designs toward being more flexible and modular. The word *SOLID* is an acronym, where each letter represents one of five principles whose names begin with that letter. These principles existed long before they were known by this name. They have proven helpful in my experience, and it is worth understanding the benefits each one brings and how we can apply them to our code. To do this, we will use a running code example in this chapter. It is a simple program that draws shapes of various kinds using simple **American Standard Code for Information Interchange** (**ASCII**) art on a console.

Before we start, let's think about the best *order* to learn these five principles. The acronym *SOLID* is easy to say, but it isn't the easiest way to learn the principles. Some principles build on others. Experience shows that some are used more than others, especially when doing TDD. For this reason, we're going to review the principles in the order *SDLOI*. It doesn't sound as good, as I'm sure you will agree, but it makes a better order of learning.

Originally, the SOLID principles were conceived as patterns that applied to classes in **object-oriented programming** (**OOP**), but they are more general-purpose than that. They equally apply to individual methods in a class as well as the class itself. They also apply to the design of microservice interconnections and function design in functional programming. We will be seeing examples applied at both the class level and the method level in this chapter.

In this chapter, we're going to cover the following main topics:

- Test guide–we drive the design
- **Single Responsibility Principle (SRP)**–simple building blocks
- **Dependency Inversion Principle (DIP)**–hiding irrelevant details
- **Liskov Substitution Principle (LSP)**–swappable objects
- **Open-Closed Principle (OCP)**–extensible design
- **Interface Segregation Principle (ISP)**–effective interfaces

Technical requirements

The code for this chapter can be found at `https://github.com/PacktPublishing/Test-Driven-Development-with-Java/tree/main/chapter07`. A running example of code that draws shapes using all five SOLID principles is provided.

Test guide – we drive the design

In *Chapter 5, Writing Our First Test*, we wrote our first test. To do that, we ran through a number of design decisions. Let's review that initial test code and list all the design decisions we had to make, as follows:

```java
@Test
public void oneIncorrectLetter() {
    var word = new Word("A");

    var score = word.guess("Z");

    assertThat( score.letter(0) ).isEqualTo(Letter.INCORRECT);
}
```

We decided on the following:

- What to test
- What to call the test
- What to call the method under test
- Which class to put that method on
- The signature of that method

- The constructor signature of the class

- Which other objects should collaborate

- The method signatures involved in that collaboration

- What form the output of this method will take

- How to access that output and assert that it worked

These are all design decisions that our human minds must make. TDD leaves us very much hands-on when it comes to designing our code and deciding how it should be implemented. To be honest, I am happy about that. Designing is rewarding and TDD provides helpful scaffolding rather than a prescriptive approach. TDD acts as a guide to remind us to make these design decisions early. It also provides a way to document these decisions as test code. Nothing more, but equally, nothing less.

It can be helpful to use techniques such as pair programming or mobbing (also known as ensemble programming) as we make these decisions—then, we add more experience and more ideas to our solution. Working alone, we simply have to take the best decisions we can, based on our own experience.

The critical point to get across here is that TDD does not and *cannot* make these decisions for us. We must make them. As such, it is useful to have some guidelines to steer us toward better designs. A set of five design principles known as the **SOLID principles** are helpful. SOLID is an acronym for the following five principles:

- SRP

- OCP

- LSP

- ISP

- DIP

In the following sections, we will learn what these principles are and how they help us write well-engineered code and tests. We will start with SRP, which is arguably the most foundational principle of any style of program design.

SRP – simple building blocks

In this section, we will examine the first principle, known as SRP. We will use a single code example throughout all sections. This will clarify how each principle is applied to an **object-oriented** (**OO**) design. We're going to look at a classic example of OO design: drawing shapes. The following diagram is an overview of the design in **Unified Modeling Language** (**UML**), describing the code presented in the chapter:

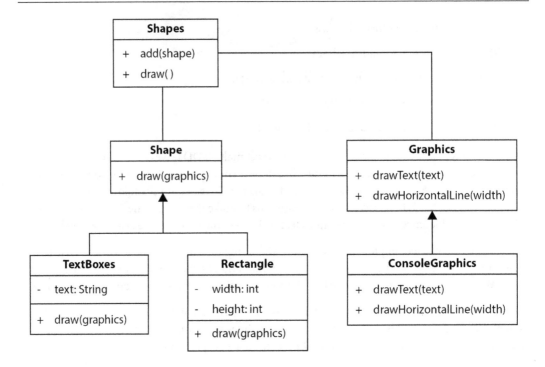

Figure 7.1 – UML diagram for shapes code

This diagram shows an overview of the Java code available in the GitHub folder for this chapter. We'll be using specific parts of the code to illustrate how each of the SOLID principles has been used to create this design.

> **UML diagrams**
>
> UML was created in 1995 by Grady Booch, Ivar Jacobson, and James Rumbaugh. UML is a way of visualizing OO designs at a high level. The preceding diagram is a UML class diagram. UML offers many other kinds of useful diagrams. You can learn more at https://www.packtpub.com/product/uml-2-0-in-action-a-project-based-tutorial/9781904811558.

SRP guides us to break code down into pieces that encapsulate a single aspect of our solution. Maybe that is a technical aspect in nature—such as reading a database table—or maybe it is a business rule. Either way, we split different aspects into different pieces of code. Each piece of code is responsible for a single detail, which is where the name *SRP* comes from. Another way of looking at this is that a piece of code should only ever have *one reason to change*. Let's examine why this is an advantage in the following sections.

Too many responsibilities make code harder to work with

A common programming mistake is to combine too many responsibilities into a single piece of code. If we have a class that can generate **Hypertext Markup Language (HTML)**, execute a business rule, and fetch data from a database table, that class will have three reasons to change. Any time a change in one of these areas is necessary, we will risk making a code change that breaks the other two aspects. The technical term for this is that the code is **highly coupled.** This leads to changes in one area rippling out and affecting other areas.

We can visualize this as code block **A** in the following diagram:

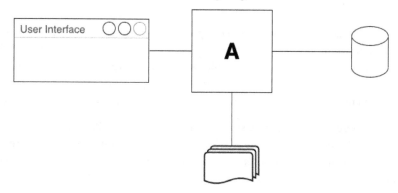

Figure 7.2 – Single component: multiple reasons to change

Block **A** deals with three things, so a change to any of them implies a change in **A**. To improve this, we apply SRP and separate out the code responsible for creating HTML, applying business rules, and accessing the database. Each of those three code blocks—**A**, **B**, and **C**—now only has one reason to change. Changing any single code block should not result in changes rippling out to the other blocks.

We can visualize this in the following diagram:

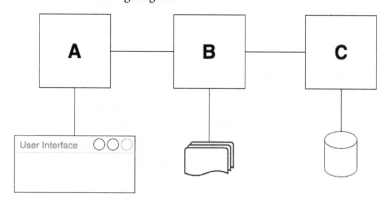

Figure 7.3 – Multiple components: one reason to change

Each code block deals with one thing and has only one reason to change. We can see that SRP works to limit the scope of future code changes. It also makes it easier to find code in a large code base, as it is logically organized.

Applying SRP gives other benefits, as follows:

- Ability to reuse code
- Simplified future maintenance

Ability to reuse code

Reusing code has been a goal of software engineering for a long time. Creating software from scratch takes time, costs money, and prevents a software engineer from doing something else. It makes sense that if we create something that is generally useful, we use it again wherever possible. The barrier to this happens when we have created large, application-specific pieces of software. The fact that they are highly specialized means they can only be used in their original context.

By creating smaller, more general-purpose software components, we will be able to use those again in different contexts. The smaller the scope of what the component aims to do, the more likely it is that we can reuse it without modification. If we have a small function or class that does one thing, it becomes easy to reuse that across our code base. It may even end up as part of a framework or library that we can reuse across multiple projects.

SRP does not guarantee that code will be reusable, but it does aim to reduce the scope of what any piece of code does. This way of thinking about code as a series of building blocks where each one does a small part of the overall task is more likely to result in reusable components.

Simplified future maintenance

As we write code, we're aware that we are not just writing to solve a problem now, but also writing code that might be revisited in the future. This might be done by other people in the team or maybe by ourselves. We want to make this future work as simple as possible. To achieve this, we need to keep our code well-engineered—making it safe and easy to work with later.

Duplicated code is a problem for maintenance—it complicates future code changes. If we copy and paste a section of code three times, let's say, it seems quite obvious to us at the time what we are doing. We have one concept that needs to happen three times, so we paste it three times. But when it comes time to read the code again, that thought process has been lost. It just reads as three unrelated pieces of code. *We lose engineering information by copy and paste.* We will need to reverse-engineer that code to work out that there are three places where we need to change it.

Counter-example – shapes code that violates SRP

To see the value of applying SRP, let's consider a piece of code that doesn't use it. The following code snippet has a list of shapes that all get drawn when we call the draw() method:

```
public class Shapes {
    private final List<Shape> allShapes = new ArrayList<>();

    public void add(Shape s) {
        allShapes.add(s);
    }

    public void draw(Graphics g) {
        for (Shape s : allShapes) {
            switch (s.getType()) {
                case "textbox":
                    var t = (TextBox) s;
                    g.drawText(t.getText());
                    break;

                case "rectangle":
                    var r = (Rectangle) s;
                    for (int row = 0;
                            row < r.getHeight();
                            row++) {
                        g.drawLine(0, r.getWidth());
                    }
            }
        }
    }
}
```

We can see that this code has four responsibilities, as follows:

- Managing the list of shapes with the add() method
- Drawing all the shapes in the list with the draw() method
- Knowing every type of shape in the switch statement
- Has implementation details for drawing each shape type in the case statements

If we want to add a new type of shape—triangle, for example—then we'll need to change this code. This will make it longer, as we need to add details about how to draw the shape inside a new `case` statement. This makes the code harder to read. The class will also have to have new tests.

Can we change this code to make adding a new type of shape easier? Certainly. Let's apply SRP and refactor.

Applying SRP to simplify future maintenance

We will refactor this code to apply SRP, taking small steps. The first thing to do is to move that knowledge of how to draw each type of shape out of this class, as follows:

```java
package shapes;

import java.util.ArrayList;
import java.util.List;

public class Shapes {
    private final List<Shape> allShapes = new ArrayList<>();

    public void add(Shape s) {
        allShapes.add(s);
    }

    public void draw(Graphics g) {
        for (Shape s : allShapes) {
            switch (s.getType()) {
                case "textbox":
                    var t = (TextBox) s;
                    t.draw(g);
                    break;

                case "rectangle":
                    var r = (Rectangle) s;
                    r.draw(g);
            }
        }
    }
}
```

The code that used to be in the `case` statement blocks has been moved into the shape classes. Let's look at the changes in the `Rectangle` class as one example—you can see what's changed in the following code snippet:

```
public class Rectangle {
    private final int width;
    private final int height;

    public Rectangle(int width, int height){
        this.width = width;
        this.height = height;
    }

    public void draw(Graphics g) {
        for (int row=0; row < height; row++) {
            g.drawHorizontalLine(width);
        }
    }
}
```

We can see how the `Rectangle` class now has the single responsibility of knowing how to draw a rectangle. It does nothing else. The one and only reason it will have to change is if we need to change how a rectangle is drawn. This is unlikely, meaning that we now have a *stable abstraction*. In other words, the `Rectangle` class is a building block we can rely on. It is unlikely to change.

If we examine our refactored `Shapes` class, we see that it too has improved. It has one responsibility less because we moved that out into the `TextBox` and `Rectangle` classes. It is simpler to read already, and simpler to test.

> **SRP**
> Do one thing and do it well. Have only one reason for a code block to change.

More improvements can be made. We see that the `Shapes` class retains its `switch` statement and that every `case` statement looks duplicated. They all do the same thing, which is to call a `draw()` method on a shape class. We can improve this by replacing the `switch` statement entirely—but that will have to wait until the next section, where we introduce the DIP.

Before we do that, let's think about how SRP applies to our test code itself.

Organizing tests to have a single responsibility

SRP also helps us to organize our *tests*. Each test should test only one thing. Perhaps this would be a single happy path or a single boundary condition. This makes it simpler to localize any faults. We find the test that failed, and because it concerns only a single aspect of our code, it is easy to find the code where the defect must be. The recommendation to only have a single assertion for each test flows naturally from this.

> **Separating tests with different configurations**
>
> Sometimes, a group of objects can be arranged to collaborate in multiple different ways. The tests for this group are often better if we write a single test per configuration. We end up with multiple smaller tests that are easier to work with.
>
> This is an example of applying SRP to each configuration of that group of objects and capturing that by writing one test for each specific configuration.

We've seen how SRP helps us create simple building blocks for our code that are simpler to test and easier to work with. The next powerful SOLID principle to look at is DIP. This is a very powerful tool for managing complexity.

DIP – hiding irrelevant details

In this section, we will learn how the DIP allows us to split code into separate components that can change independently of each other. We will then see how this naturally leads to the OCP part of SOLID.

Dependency inversion (**DI**) means that we write code to depend on abstractions, not details. The opposite of this is having two code blocks, one that depends on the detailed implementation of the other. Changes to one block will cause changes to another. To see what this problem looks like in practice, let's review a counter-example. The following code snippet begins where we left off with the Shapes class after applying SRP to it:

```java
package shapes;

import java.util.ArrayList;
import java.util.List;

public class Shapes {
```

```
private final List<Shape> allShapes = new ArrayList<>();

public void add(Shape s) {
    allShapes.add(s);
}


public void draw(Graphics g) {
    for (Shape s : allShapes) {
        switch (s.getType()) {
            case "textbox":
                var t = (TextBox) s;
                t.draw(g);
                break;

            case "rectangle":
                var r = (Rectangle) s;
                r.draw(g);
        }
    }
}
}
```

This code does work well to maintain a list of Shape objects and draw them. The problem is that it knows too much about the types of shapes it is supposed to draw. The draw() method features a **switch-on-type** of object that you can see. That means that if anything changes about which types of shapes should be drawn, then this code must also change. If we want to add a new Shape to the system, then we have to modify this switch statement and the associated TDD test code.

The technical term for one class knowing about another is that a **dependency** exists between them. The Shapes class *depends on* the TextBox and Rectangle classes. We can represent that visually in the following UML class diagram:

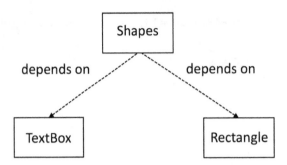

Figure 7.4 – Depending on the details

We can see that the `Shapes` class depends directly on the detail of the `Rectangle` and `TextBox` classes. This is shown by the direction of the arrows in the UML class diagram. Having these dependencies makes working with the `Shapes` class more difficult for the following reasons:

- We have to change the `Shapes` class to add a new kind of shape

- Any changes in the concrete classes such as `Rectangle` will cause this code to change

- The `Shapes` class will get longer and less easy to read

- We will end up with more test cases

- Each test case will be coupled to concrete classes such as `Rectangle`

This is a very procedural approach to creating a class that deals with multiple kinds of shapes. It violates SRP by doing too much and knowing too much detail about each kind of shape object. The `Shapes` class depends on the details of concrete classes such as `Rectangle` and `TextBox`, which directly causes the aforementioned problems.

Thankfully, there is a better way. We can use the power of an interface to improve this, by making it so that the `Shapes` class does *not* depend on those details. This is called DI. Let's see what that looks like next.

Applying DI to the shapes code

We can improve the shapes code by applying the **Dependency Inversion Principle (DIP)** described in the previous chapter. Let's add a `draw()` method to our `Shape` interface, as follows:

```
package shapes;

public interface Shape {
    void draw(Graphics g);
}
```

This interface is our abstraction of the single responsibility that each shape has. Each shape must know how to draw itself when we call the `draw()` method. The next step is to make our concrete shape classes implement this interface.

Let's take the `Rectangle` class as an example. You can see this here:

```
public class Rectangle implements Shape {
    private final int width;
    private final int height;

    public Rectangle(int width, int height){
        this.width = width;
        this.height = height;
    }

    @Override
    public void draw(Graphics g) {
        for (int row=0; row < height; row++) {
            g.drawHorizontalLine(width);
        }
    }
}
```

We've now introduced the OO concept of polymorphism into our shape classes. This breaks the dependency that the `Shapes` class has on knowing about the `Rectangle` and `TextBox` classes. All that the `Shapes` class now depends on is the `Shape` interface. It no longer needs to know the type of each shape.

We can refactor the `Shapes` class to look like this:

```
public class Shapes {
    private final List<Shape> all = new ArrayList<>();

    public void add(Shape s) {
        all.add(s);
    }

    public void draw(Graphics graphics) {
        all.forEach(shape->shape.draw(graphics));
```

```
        }
    }
```

This refactoring has completely removed the switch statement and the getType() method, making the code much simpler to understand and test. If we add a new kind of shape, the Shapes class *no longer needs to change*. We have broken that dependency on knowing the details of shape classes.

One minor refactor moves the Graphics parameter we pass into the draw() method into a field, initialized in the constructor, as illustrated in the following code snippet:

```
public class Shapes {
    private final List<Shape> all = new ArrayList<>();
    private final Graphics graphics;

    public Shapes(Graphics graphics) {
        this.graphics = graphics;
    }

    public void add(Shape s) {
        all.add(s);
    }

    public void draw() {
        all.forEach(shape->shape.draw(graphics));
    }
}
```

This is DIP at work. We've created an abstraction in the Shape interface. The Shapes class is a consumer of this abstraction. The classes implementing that interface are providers. Both sets of classes depend only on the abstraction; they do not depend on details inside each other. There are no references to the Rectangle class in the Shapes class, and there are no references to the Shapes inside the Rectangle class. We can see this inversion of dependencies visualized in the following UML class diagram—see how the direction of the dependency arrows has changed compared to *Figure 7.4*:

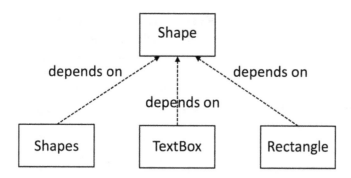

Figure 7.5 – Inverting dependencies

In this version of the UML diagram, the arrows describing the dependencies between classes point the opposite way. *The dependencies have been inverted*—hence, the name of this principle. Our Shapes class now depends on our abstraction, the Shape interface. So do all the Rectangle class and TextBox class concrete implementations. We have inverted the dependency graph and turned the arrows upside down. DI fully decouples classes from each other and, as such, is very powerful. We will see how this leads to a key technique for TDD testing when we look at *Chapter 8, Test Doubles – Stubs and Mocks*.

> **DIP**
>
> Make code depend on abstractions and not on details.

We've seen how DIP is a major tool we can use to simplify our code. It allows us to write code that deals with an interface, and then use that code with any concrete class that implements that interface. This begs a question: can we write a class that implements an interface but will not work correctly? That's the subject of our next section.

LSP – swappable objects

Turing Award winner Barbara Liskov is the creator of a rule concerning inheritance that is now commonly known as LSP. It was brought about by a question in OOP: if we can extend a class and use it in place of the class we extended, how can we be sure the new class will not break things?

We've seen in the previous section on DIP how we can use any class that implements an interface in place of the interface itself. We also saw how those classes can provide any implementation they like for that method. The interface itself provides no guarantees at all about what might lurk inside that implementation code.

There is, of course, a bad side to this—which LSP aims to avoid. Let's explain this by looking at a counter-example in code. Suppose we made a new class that implemented interface Shape, such as this one (Warning: Do *NOT* run the code that follows in the MaliciousShape class!):

```java
public class MaliciousShape implements Shape {
    @Override
    public void draw(Graphics g) {
        try {
            String[] deleteEverything = {"rm", "-Rf", "*"};
            Runtime.getRuntime().exec(deleteEverything,null);

            g.drawText("Nothing to see here...");
        } catch (Exception ex) {
            // No action
        }
    }
}
```

Notice anything a little odd about that new class? It contains a Unix command to remove all our files! This is not what we are expecting when we call the draw() method on a shape object. Due to permissions failures, it might not be able to delete anything, but it's an example of what can go wrong.

An interface in Java can only protect the *syntax* of method calls we expect. It cannot enforce any *semantics*. The problem with the preceding MaliciousShape class is that it does not respect the intent behind the interface.

LSP guides us to avoid this error. In other words, LSP states that any class that implements an interface or extends another class must handle all the input combinations that the original class/interface could. It must provide the expected outputs, it must not ignore valid inputs, and it must not produce completely unexpected and undesired behavior. Classes written like this are safe to use through a reference to their interface. The problem with our MaliciousShape class is that it was not compatible with LSP—it added some extra totally unexpected and unwanted behavior.

> **LSP formal definition**
>
> American computer scientist Barbara Liskov came up with a formal definition: If $p(x)$ is a property provable about objects x of type T, then $p(y)$ should be true for objects y of type S where S is a subtype of T.

Reviewing LSP usage in the shapes code

The classes that implement Shape all conform to LSP. This is clear in the TextBox class, as we can see here:

```
public class TextBox implements Shape {
    private final String text;

    public TextBox(String text) {
        this.text = text;
    }

    @Override
    public void draw(Graphics g) {
        g.drawText(text);
    }
}
```

The preceding code clearly can handle drawing any valid text provided to its constructor. It also provides no surprises. It draws the text, using primitives from the Graphics class, and does nothing else.

Other examples of LSP compliance can be seen in the following classes:

- Rectangle
- Triangle

LSP

A code block can be safely swapped for another if it can handle the full range of inputs and provide (at least) all expected outputs, with no undesired side effects.

There are some surprising violations of LSP. Perhaps the classic one for the shapes code example is about adding a Square class. In mathematics, a square is a kind of rectangle, with the extra constraint that its height and width are equal. In Java code, should we make the Square class extend the Rectangle class? How about the Rectangle class extending Square?

Let's apply LSP to decide. We will imagine some code that expects a Rectangle class so that it can change its height, but not its width. If we passed a Square class to that code, would it work properly? The answer is no. You would then have a square with unequal width and height. This *fails* LSP.

The point of LSP is about making classes properly conform to interfaces. In the next section, we'll look at OCP, which is closely related to DI.

OCP – extensible design

In this section, we'll see how OCP helps us write code that we can add new features to, without changing the code itself. This does sound like an impossibility at first, but it flows naturally from DIP combined with LSP.

OCP results in code that is open to extension but closed to modification. We saw this idea at work when we looked at DIP. Let's review the code refactoring we did in the light of OCP.

Let's start with the original code for the Shapes class, as follows:

```java
public class Shapes {
    private final List<Shape> allShapes = new ArrayList<>();

    public void add(Shape s) {
        allShapes.add(s);
    }

    public void draw(Graphics g) {
        for (Shape s : allShapes) {
            switch (s.getType()) {
                case "textbox":
                    var t = (TextBox) s;
                    g.drawText(t.getText());
                    break;

                case "rectangle":
                    var r = (Rectangle) s;
                    for (int row = 0;
                        row < r.getHeight();
                        row++) {
                        g.drawLine(0, r.getWidth());
                    }
            }
        }
    }
}
```

Adding a new type of shape requires modification of the code inside the draw () method. We will be adding a new case statement in to support our new shape.

Modifying existing code has several disadvantages, as set out here:

- We invalidate prior testing. This is now different code than we had tested.

- We might introduce an error that breaks some of the existing support for shapes.

- The code will become longer and more difficult to read.

- We might have several developers add shapes at the same time and get a merge conflict when we combine their work.

By applying DIP and refactoring the code, we ended up with this:

```
public class Shapes {
    private final List<Shape> all = new ArrayList<>();
    private final Graphics graphics;

    public Shapes(Graphics graphics) {
        this.graphics = graphics;
    }

    public void add(Shape s) {
        all.add(s);
    }

    public void draw() {
        all.forEach(shape->shape.draw(graphics));
    }
}
```

We can now see that adding a new type of shape *does not need modification to this code*. This is an example of OCP at work. The Shapes class is *open* to having new kinds of shapes defined, but it is *closed* against the need for modification when that new shape is added. This also means that any tests relating to the Shapes class will remain unchanged, as there is no difference in behavior for this class. That is a powerful advantage.

OCP relies on DI to work. It is more or less a restatement of a consequence of applying DIP. It also provides us with a technique to support *swappable behavior*. We can use DIP and OCP to create plugin systems.

Adding a new type of shape

To see how this works in practice, let's create a new type of shape, the RightArrow class, as follows:

```java
public class RightArrow implements Shape {
  public void draw(Graphics g) {
    g.drawText( "    \" );
    g.drawText( "-----" );
    g.drawText( "    /" );
  }
}
```

The RightArrow class implements the Shape interface and defines a draw() method. To demonstrate that nothing in the Shapes class needs to change in order to use this, let's review some code that uses both the Shapes and our new class, RightArrow, as follows:

```java
package shapes;

public class ShapesExample {
    public static void main(String[] args) {
        new ShapesExample().run();
    }

    private void run() {
        Graphics console = new ConsoleGraphics();
        var shapes = new Shapes(console);

        shapes.add(new TextBox("Hello!"));
        shapes.add(new Rectangle(32,1));
        shapes.add(new RightArrow());

        shapes.draw();
    }
}
```

We see that the Shapes class is being used in a completely normal way, without change. In fact, the only change needed to use our new RightArrow class is to create an object instance and pass it to the add() method of shapes.

> **OCP**
>
> Make code open for new behaviors, but closed for modifications.

The power of OCP should now be clear. We can extend the capabilities of our code and keep changes limited. We greatly reduce the risk of breaking code that is already working, as we no longer need to change that code. OCP is a great way to manage complexity. In the next section, we'll look at the remaining SOLID principle: ISP.

ISP – effective interfaces

In this section, we will look at a principle that helps us write effective interfaces. It is known as ISP.

ISP advises us to keep our interfaces small and dedicated to achieving a single responsibility. By small interfaces, we mean having as few methods as possible on any single interface. These methods should all relate to some common theme.

We can see that this principle is really just SRP in another form. We are saying that an effective interface should describe a single responsibility. It should cover one abstraction, not several. The methods on the interface should strongly relate to each other and also to that single abstraction.

If we need more abstractions, then we use more interfaces. We keep each abstraction in its own separate interface, which is where the term *interface segregation* comes from —we keep different abstractions apart.

The related **code smell** to this is a large interface that covers several different topics in one. We could imagine an interface having hundreds of methods in little groups—some relating to file management, some about editing documents, and some about printing documents. Such interfaces quickly become difficult to work with. ISP suggests that we improve this by splitting the interface into several smaller ones. This split would preserve the groups of methods—so, you might see interfaces for file management, editing, and printing, with relevant methods under each. We have made our code simpler to understand by splitting apart these separate abstractions.

Reviewing ISP usage in the shapes code

The most noticeable use of ISP is in the Shape interface, as illustrated here:

```
interface Shape {
   void draw(Graphics g);
}
```

This interface clearly has a single focus. It is an interface with a very narrow focus, so much so that only one method needs to be specified: draw(). There is no confusion arising from other mixed-in concepts here and no unnecessary methods. That single method is both necessary and sufficient. The other major example is in the `Graphics` interface, as shown here:

```
public interface Graphics {
    void drawText(String text);
    void drawHorizontalLine(int width);
}
```

The `Graphics` interface contains only methods related to drawing graphics primitives on screen. It has two methods—`drawText` to display a text string, and `drawHorizontalLine` to draw a line in a horizontal direction. As these methods are strongly related—known technically as exhibiting **high cohesion**—and few in number, ISP is satisfied. This is an effective abstraction over the graphics drawing subsystem, tailored to our purposes.

For completeness, we can implement this interface in a number of ways. The example in GitHub uses a simple text console implementation:

```
public class ConsoleGraphics implements Graphics {
    @Override
    public void drawText(String text) {
        print(text);
    }

    @Override
    public void drawHorizontalLine(int width) {
        var rowText = new StringBuilder();

        for (int i = 0; i < width; i++) {
            rowText.append('X');
        }

        print(rowText.toString());
    }

    private void print(String text) {
        System.out.println(text);
    }
}
```

That implementation is also LSP-compliant—it can be used wherever the `Graphics` interface is expected.

> **ISP**
>
> Keep interfaces small and strongly related to a single idea.

We've now covered all five of the SOLID principles and shown how they have been applied to the shapes code. They have guided the design toward compact code, having a well-engineered structure to assist future maintainers. We know how to incorporate these principles into our own code to gain similar benefits.

Summary

In this chapter, we've looked at simple explanations of how the SOLID principles help us design both our production code and our tests. We've worked through an example design that uses all five SOLID principles. In future work, we can apply SRP to help us understand our design and limit the rework involved in future changes. We can apply DIP to split up our code into independent small pieces, leaving each piece to hide some of the details of our overall program, creating a divide-and-conquer effect. Using LSP, we can create objects that can be safely and easily swapped. OCP helps us design software that is simple to add functionality to. ISP will keep our interfaces small and easy to understand.

The next chapter puts these principles to use to solve a problem in testing—how do we test the collaborations between our objects?

Questions and answers

1. Do the SOLID principles only apply to OO code?

   No. Although originally applied to an OO context, they have uses in both functional programming and microservice design. SRP is almost universally useful—sticking to one main focus is helpful for anything, even paragraphs of documentation. SRP thinking also helps us write a pure function that does only one thing and a test that does only one thing. DIP and OCP are easily done in functional contexts by passing in the dependency as a pure function, as we do with Java lambdas. SOLID as a whole gives a set of goals for managing coupling and cohesion among any kind of software components.

2. Do we have to use SOLID principles with TDD?

   No. TDD works by defining the outcomes and public interface of a software component. How we implement that component is irrelevant to a TDD test, but using principles such as SRP and DIP makes it much easier to write tests against that code by giving us the test access points we need.

3. Are SOLID principles the only ones we should use?

    No. We should use every technique at our disposal.

    The SOLID principles make a great starting point in shaping your code and we should take advantage of them, but there are many other valid techniques to design software. The whole catalog of design patterns, the excellent system of **General Responsibility Assignment Software Patterns (GRASP)** by Craig Larman, the idea of information hiding by David L. Parnas, and the ideas of coupling and cohesion all apply. We should use any and every technique we know—or can learn about—to serve our goal of making software that is easy to read and safe to change.

4. If we do not use the SOLID principles, can we still do TDD?

    Yes—very much so. TDD concerns itself with testing the behavior of code, not the details of how it is implemented. SOLID principles simply help us create OO designs that are robust and simpler to test.

5. How does SRP relate to ISP?

    ISP guides us to prefer many shorter interfaces over one large interface. Each of the shorter interfaces should relate to one single aspect of what a class should provide. This is usually some kind of role, or perhaps a subsystem. ISP can be thought of as making sure our interfaces each apply the SRP and do only one thing—well.

6. How does OCP relate to DIP and LSP?

    OCP guides us to create software components that can have new capabilities added without changing the component itself. This is done by using a plugin design. The component will allow separate classes to be plugged in providing the new capabilities. The way to do this is to create an abstraction of what a plugin should do in an interface—DIP. Then, create concrete plugin implementations of this conforming to LSP. After that, we can inject these new plugins into our component. OCP relies on DIP and LSP to work.

8

Test Doubles – Stubs and Mocks

In this chapter, we're going to solve a common testing challenge. How do you test an object that depends on another object? What do we do if that collaborator is difficult to set up with test data? Several techniques are available to help us with this and they build on the SOLID principles we learned previously. We can use the idea of dependency injection to enable us to replace collaborating objects with ones specially written to help us write our test.

These new objects are called test doubles, and we will learn about two important kinds of test double in this chapter. We will learn when to apply each kind of test double and then learn two ways of creating them in Java – both by writing the code ourselves and by using the popular library Mockito. By the end of the chapter, we will have techniques that allow us to write tests for objects where it is difficult or impossible to test them with the real collaborating objects in place. This allows us to use TDD with complex systems.

In this chapter, we're going to cover the following main topics:

- The problems of testing collaborators
- The purpose of test doubles
- Using stubs for pre-canned results
- Using mocks to verify interactions
- Understanding when test doubles are appropriate
- Working with Mockito – a popular mocking library
- Driving error handling code using stubs
- Testing an error condition in Wordz

Technical requirements

The code for this chapter can be found at https://github.com/PacktPublishing/Test-Driven-Development-with-Java/tree/main/chapter08.

The problems collaborators present for testing

In this section, we will understand the challenges that arise as we grow our software into a larger code base. We will review what is meant by a collaborating object, then we will take a look at two examples of collaborations that are challenging to test.

As we grow our software system, we will soon outgrow what can go in a single class (or function, for that matter). We will split our code into multiple parts. If we pick a single object as our subject under test, any other object that it depends on is a collaborator. Our TDD tests must account for the presence of these collaborators. Sometimes, this is straightforward, as we've seen in earlier chapters.

Unfortunately, things aren't always that simple. Some collaborations make tests difficult – or impossible – to write. These kinds of collaborators introduce either unrepeatable behaviors that we must contend with or present errors that are difficult to trigger.

Let's review these challenges with some short examples. We'll start with a common problem: a collaborator that exhibits unrepeatable behavior.

The challenges of testing unrepeatable behavior

We've learned that the basic steps of a TDD test are Arrange, Act, and Assert. We ask the object to act and then assert that an expected outcome happens. But what happens when that outcome is unpredictable?

To illustrate, let's review a class that rolls a die and presents a text string to say what we rolled:

```java
package examples;

import java.util.random.RandomGenerator;

public class DiceRoll {

    private final int NUMBER_OF_SIDES = 6;
    private final RandomGenerator rnd =
                        RandomGenerator.getDefault();

    public String asText() {
        int rolled = rnd.nextInt(NUMBER_OF_SIDES) + 1;

        return String.format("You rolled a %d", rolled);
    }
}
```

This is simple enough code, with only a handful of executable lines in it. Sadly, *simple to write is not always simple to test*. How would we write a test for this? Specifically – how would we write the assert? In previous tests, we've always known exactly what to expect in the assertion. Here, the assertion will be some fixed text plus a random number. We don't know in advance what that random number will be.

The challenges of testing error handling

Testing code that handles error conditions is another challenge. The difficulty here lies not in asserting that the error was handled, but rather the challenge is how to trigger that error to happen inside the collaborating object.

To illustrate, let's imagine a code to warn us when the battery in our portable device is getting low:

```
public class BatteryMonitor {
    public void warnWhenBatteryPowerLow() {
        if (DeviceApi.getBatteryPercentage() < 10) {
            System.out.println("Warning - Battery low");
        }
    }
}
```

The preceding code in `BatteryMonitor` features a `DeviceApi` class, which is a library class that lets us read how much battery we have left on our phone. It provides a static method to do this, called `getBatteryPercentage()`. This will return an integer in the range *0* to *100* percent. The code that we want to write a TDD test for calls `getBatteryPercentage()` and will display a warning message if it is less than *10* percent. But there's a problem writing this test: how can we force the `getBatteryPercentage()` method to return a number less than 10 as part of our Arrange step? Would we discharge the battery somehow? How would we do this?

`BatteryMonitor` provides an example of code that collaborates with another object, where it is impossible to force a known response from that collaborator. We have no way to change the value that `getBatteryPercentage()` will return. We would literally have to wait until the battery had discharged before this test could pass. That's not what TDD is about.

Understanding why these collaborations are challenging

When doing TDD, we want *fast* and *repeatable* tests. Any scenario that involves unpredictable behavior or requires us to control a situation that we have no control over clearly causes problems for TDD.

The best way to write tests in these cases is by eliminating the cause of the difficulty. Fortunately, a simple solution exists. We can apply the *Dependency Injection Principle* we learned about in the previous chapter, along with one new idea – the *test double*. We will review test doubles in the next section.

The purpose of test doubles

In this section, we're going to learn techniques that allow us to test these challenging collaborations. We will introduce the idea of test doubles. We will learn how to apply the SOLID principles to design code flexible enough to use these test doubles.

The challenges of the previous section are solved by using **test doubles**. A test double replaces one of the collaborating objects in our test. By design, this test double avoids the difficulties of the replaced object. Think of them as the stunt doubles in movies, replacing the real actors to help safely get an action shot.

A software test double is an object we have written specifically to be easy to use in our unit test. In the test, we inject our test double into the SUT in the Arrange step. In production code, we inject in the production object that our test double had replaced.

Let's reconsider our `DiceRoll` example earlier. How would we refactor that code to make it easier to test?

1. Create an interface that abstracts the source of random numbers:

    ```
    interface RandomNumbers {
        int nextInt(int upperBoundExclusive);
    }
    ```

2. Apply the *Dependency Inversion Principle* to `class DiceRoll` to make use of this abstraction:

    ```
    package examples;

    import java.util.random.RandomGenerator;

    public class DiceRoll {

        private final int NUMBER_OF_SIDES = 6;
        private final RandomNumbers rnd ;

        public DiceRoll( RandomNumbers r ) {
            this.rnd = r;
        }
        public String asText() {
            int rolled = rnd.nextInt(NUMBER_OF_SIDES) + 1;

            return String.format("You rolled a %d",
    ```

```
                                            rolled);
    }
}
```

We have inverted the dependency on the random number generator by replacing it with the
RandomNumbers interface. We added a constructor that allows a suitable RandomNumbers
implementation to be injected. We assign that to the rnd field. The asText() method now
calls the nextInt() method on whatever object we passed to the constructor.

3. Write a test, using a test double to replace the RandomNumbers source:

```java
package examples;

import org.junit.jupiter.api.Test;
import static org.assertj.core.api.Assertions.assertThat;

class DiceRollTest {
    @Test
    void producesMessage() {
        var stub = new StubRandomNumbers();
        var roll = new DiceRoll(stub);

        var actual = roll.asText();

        assertThat(actual).isEqualTo("You rolled a
                                        5");
    }
}
```

We see the usual Arrange, Act, and Assert sections in this test. The new idea here is class
StubRandomNumbers. Let's look at the stub code:

```java
package examples;

public class StubRandomNumbers implements RandomNumbers {

    @Override
    public int nextInt(int upperBoundExclusive) {
        return 4;   // @see https://xkcd.com/221
    }
}
```

There are a few things to notice about this stub. Firstly, it implements our RandomNumbers interface, making it an LSP-compliant substitute for that interface. This allows us to inject it into the constructor of DiceRoll, our SUT. The second most important aspect is that every call to nextInt() will *return the same number*.

By replacing the real RandomNumbers source with a stub that delivers a known value, we have made our test assertion easy to write. The stub eliminates the problem of unrepeatable values from the random generator.

We can now see how the DiceRollTest works. We supply a test double to our SUT. The test double always returns the same value. As a result, we can assert against a known outcome.

Making the production version of the code

To make class DiceRoll work properly in production, we would need to inject a genuine source of random numbers. A suitable class would be the following:

```
public class RandomlyGeneratedNumbers implements RandomNumbers
{
    private final RandomGenerator rnd =
                        RandomGenerator.getDefault();

    @Override
    public int nextInt(int upperBoundExclusive) {

        return rnd.nextInt(upperBoundExclusive);
    }
}
```

There isn't much work to do here – the preceding code simply implements the nextInt() method using the RandomGenerator library class built into Java.

We can now use this to create our production version of the code. We already changed our DiceRoll class to allow us to inject in any suitable implementation of the RandomNumbers interface. For our test code, we injected in a test double – an instance of the StubRandomNumbers class. For our production code, we will instead inject in an instance of the RandomlyGeneratedNumbers class. The production code will use that object to create real random numbers – and there will be no code changes inside the DiceRoll class. We have used the Dependency Inversion Principle to make class DiceRoll configurable by dependency injection. This means that class DiceRoll now follows the Open/Closed Principle – it is *open* to new kinds of random number generation behavior but *closed* to code changes inside the class itself.

> **Dependency inversion, dependency injection, and inversion of control**
>
> The preceding example shows these three ideas in action. *Dependency inversion* is the design technique where we create an abstraction in our code. *Dependency injection* is the runtime technique where we supply an implementation of that abstraction to code that depends on it. Together, these ideas are often termed **Inversion of Control (IoC)**. Frameworks such as Spring are sometimes called IoC containers because they provide tools to help you manage creating and injecting dependencies in an application.

The following code is an example of how we would use DiceRoll and RandomlyGeneratedNumbers in production:

```java
public class DiceRollApp {
    public static void main(String[] args) {
        new DiceRollApp().run();
    }


    private void run() {
        var rnd = new RandomlyGeneratedNumbers();
        var roll = new DiceRoll(rnd);

        System.out.println(roll.asText());
    }
}
```

You can see in the previous code that we inject an instance of the production-version RandomlyGeneratedNumbers class into the DiceRoll class. This process of creating and injecting objects is often termed **object wiring**. Frameworks such as *Spring* (https://spring.io/), *Google Guice* (https://github.com/google/guice), and the built-in *Java CDI* (https://docs.oracle.com/javaee/6/tutorial/doc/giwhl.html) provide ways to minimize the boilerplate of creating dependencies and wiring them up, using annotations.

Using DIP to swap a production object for a test double is a very powerful technique. This test double is an example of a kind of double known as a stub. We'll cover what a stub is along with when to use one in the next section.

Using stubs for pre-canned results

The previous section explained that test doubles were a kind of object that could stand in for a production object so that we could write a test more easily. In this section, we will take a closer look at that test double and generalize it.

In the preceding `DiceRoll` example, the test was simpler to write because we replaced the random number generation with a known, fixed value. Our genuine random number generator made it difficult to write an assertion, as we were never sure what the expected random number should be. Our test double was an object that instead supplied a well-known value. We can then work out the expected value for our assertion, making our test easy to write.

A test double that supplies values like this is called a **stub**. Stubs always replace an object that we cannot control with a test-only version that we can control. They always produce known data values for our code under test to consume. Graphically, a stub looks like this:

Figure 8.1 – Replacing a collaborator with a stub

In the diagram, our test class is responsible for wiring up our SUT to an appropriate stub object in the Arrange step. When the Act step asks our SUT to execute the code we want to test, that code will pull the known data values from the stub. The Assert step can be written based on the expected behavior that these known data values will cause.

It is important to note why this works. One objection to this arrangement is that we are not testing the real system. Our SUT is wired up to some object that will never be part of our production system. That is true. But this works because our test is only testing the logic within the SUT. This test is *not* testing the behavior of the dependencies themselves. Indeed, it must not attempt to do that. Testing the test double is a classic anti-pattern for unit tests.

Our SUT has used the Dependency Inversion Principle to fully isolate itself from the object the stub is standing in for. It makes no difference to the SUT how it gets its data from its collaborator. That's why this testing approach is valid.

When to use stub objects

Stubs are useful whenever our SUT uses a *pull model* of collaborating with a dependency. Some examples of when using stubs makes sense are as follows:

- **Stubbing a repository interface/database**: Using a stub instead of calling to a real database for data access code

- **Stubbing reference data sources**: Replacing properties files or web services containing reference data with stub data

- **Providing application objects to code that converts to HTML or JSON formats**: When testing code that converts to HTML or JSON, supply input data with a stub

- **Stubbing the system clock to test time-dependent behavior**: To get repeatable behavior out of a time call, stub the call with known times

- **Stubbing random number generators to create predictability**: Replace a call to a random number generator with a call to a stub

- **Stubbing authentication systems to always allow a test user to log in**: Replace calls to authentication systems with simple "login succeeded" stubs

- **Stubbing responses from a third-party web service such as a payment provider**: Replace real calls to third-party services with calls to a stub

- **Stubbing a call to an operating system command**: Replace a call to the OS to, for example, list a directory with pre-canned stub data

In this section, we have seen how using stubs allows us to control data that gets supplied to an SUT. It supports a *pull model* of fetching objects from elsewhere. But that's not the only mechanism by which objects can collaborate. Some objects use a *push model*. In this case, when we call a method on our SUT, we expect it to call another method on some other object. Our test must confirm that this method call actually took place. This is something that stubs cannot help with and needs a different approach. We will cover this approach in the next section.

Using mocks to verify interactions

In this section, we'll take a look at another important kind of test double: the mock object. Mock objects solve a slightly different problem than stub objects do, as we shall see in this section.

Mock objects are a kind of test double that *record interactions*. Unlike stubs, which supply well-known objects to the SUT, a mock will simply record interactions that the SUT has with the mock. It is the perfect tool to answer the question, "*Did the SUT call the method correctly?*" This solves the problem of *push model* interactions between the SUT and its collaborator. The SUT commands the collaborator to do something rather than requesting something from it. A mock provides a way to verify that it issued that command, along with any necessary parameters.

The following UML object diagram shows the general arrangement:

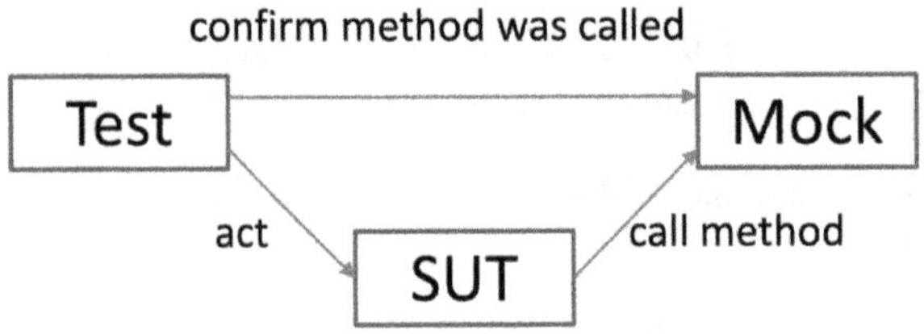

Figure 8.2 – Replace collaborator with mock

We see our test code wiring up a mock object to the SUT. The Act step will make the SUT execute code that we expect to interact with its collaborator. We have swapped out that collaborator for a mock, which will record the fact that a certain method was called on it.

Let's look at a concrete example to make this easier to understand. Suppose our SUT is expected to send an email to a user. Once again, we will use the Dependency Inversion Principle to create an abstraction of our mail server as an interface:

```
public interface MailServer {
    void sendEmail(String recipient, String subject,
                String text);
}
```

The preceding code shows a simplified interface only suitable for sending a short text email. It is good enough for our purposes. To test the SUT that called the sendEmail() method on this interface, we would write a MockMailServer class:

```
public class MockMailServer implements MailServer {
    boolean wasCalled;
    String actualRecipient;
    String actualSubject;
    String actualText;

    @Override
    public void sendEmail(String recipient, String subject,
                        String text) {
        wasCalled = true;
        actualRecipient = recipient;
        actualSubject = subject;
        actualText = text;
```

```
        }
    }
```

The preceding `MockMailServer` class implements the `MailServer` interface. It has a single responsibility – to record the fact that the `sendEmail()` method was called and to capture the actual parameter values sent to that method. It exposes these as simple fields with package-public visibility. Our test code can use these fields to form the assertion. Our test simply has to wire up this mock object to the SUT, cause the SUT to execute code that we expect to call the `sendEmail()` method, and then check that it did do that:

```
@Test
public void sendsWelcomeEmail() {
    var mailServer = new MockMailServer();
    var notifications = new UserNotifications(mailServer);

    notifications.welcomeNewUser();

    assertThat(mailServer.wasCalled).isTrue();

    assertThat(mailServer.actualRecipient)
        .isEqualTo("test@example.com");

    assertThat(mailServer.actualSubject)
        .isEqualTo("Welcome!");

    assertThat(mailServer.actualText)
        .contains("Welcome to your account");
}
```

We can see that this test wires up the mock to our SUT, then causes the SUT to execute the `welcomeNewUser()` method. We expect this method to call the `sendEmail()` method on the `MailServer` object. Then, we need to write assertions to confirm that call was made with the correct parameter values passed. We're using the idea of four assert statements logically here and testing one idea – effectively a single assert.

The power of mock objects is that we can *record interactions* with objects that are difficult to control. In the case of a mail server, such as the one seen in the preceding code block, we would not want to be sending actual emails to anybody. We also would not want to write a test that waited around monitoring the mailbox of a test user. Not only is this slow and can be unreliable, but it is also not what we intend to test. The SUT only has the responsibility of making the call to `sendEmail()` – what happens after that is out of the scope of the SUT. It is, therefore, out of scope for this test.

As in the previous examples with other test doubles, the fact that we have used the **Dependency Inversion Principle** means our production code is easy enough to create. We simply need to create an implementation of `MailServer` that uses the SMTP protocol to talk to a real mail server. We would most likely search for a library class that does that for us already, then we would need to make a very simple adapter object that binds that library code to our interface.

This section has covered two common kinds of test double, stubs, and mocks. But test doubles are not always appropriate to use. In the next section, we'll discuss some issues to be aware of when using test doubles.

Understanding when test doubles are appropriate

Mock objects are a useful kind of test double, as we have seen. But they are not always the right approach. There are some situations where we should actively avoid using mocks. These situations include over-using mocks, using mocks for code you don't own, and mocking value objects. We'll look at these situations next. Then, we'll recap with general advice for where mocks are typically useful. Let's start by considering the problems caused when we overuse mock objects.

Avoiding the overuse of mock objects

At a first glance, using mock objects seems to solve a number of problems for us. Yet if used without care, we can end up with very poor-quality tests. To understand why, let's go back to our basic definition of a TDD test. It is a test that verifies *behaviors* and is independent of *implementations*. If we use a mock object to stand in for a genuine abstraction, then we are complying with that.

The potential problem happens because it is all too easy to create a mock object for an implementation detail, not an abstraction. If we do this, we end up locking our code into a specific implementation and structure. Once a test is coupled to a specific implementation detail, then changing that implementation requires a change to the test. If the new implementation has the same outcomes as the old one, the test really should still pass. Tests that depend on specific implementation details or code structures actively impede refactoring and adding new features.

Don't mock code you don't own

Another area where mocks should not be used is as a stand-in for a concrete class written outside of your team. Suppose we are using a class called `PdfGenerator` from a library to create a PDF document. Our code would call methods on the `PdfGenerator` class. We might think it would be easy to test our code if we use a mock object to stand in for the `PdfGenerator` class.

This approach has a problem that will only arise in the future. The class in the external library will quite likely change. Let's say that the `PdfGenerator` class removes one of the methods our code is calling. We will be forced to update the library version at some point as part of our security policy if nothing else. When we pull in the new version, our code will no longer compile against this changed class – *but our tests will still pass because the mock object still has the old method in it.* This is a subtle trap that we have laid for future maintainers of the code. It is best avoided. A reasonable approach is to wrap the third-party library, and ideally place it behind an interface to invert the dependency on it, isolating it fully.

Don't mock value objects

A **value object** is an object that has no specific identity, it is defined only by the data it contains. Some examples would include an integer or a string object. We consider two strings to be the same if they contain the same text. They might be two separate string objects in memory, but if they hold the same value, we consider them to be equal.

The clue that something is a value object in Java is the presence of a customized `equals()` and `hashCode()` method. By default, Java compares the equality of two objects using their identity – it checks that two object references are referring to the same object instance in memory. We must override the `equals()` and `hashCode()` methods to provide the correct behavior for value objects, based on their content.

A value object is a simple thing. It may have some complex behaviors inside its methods but, in principle, value objects should be easy to create. There is no benefit in creating a mock object to stand in for one of these value objects. Instead, create the value object and use it in your test.

You can't mock without dependency injection

Test doubles can only be used where we can inject them. This is not always possible. If the code we want to test creates a concrete class using the new keyword, then we cannot replace it with a double:

```
package examples;

public class UserGreeting {

    private final UserProfiles profiles
        = new UserProfilesPostgres();

    public String formatGreeting(UserId id) {
        return String.format("Hello and welcome, %s",
            profiles.fetchNicknameFor(id));
    }
}
```

We see that the `profiles` field has been initialized using a concrete class `UserProfilesPostgres()`. There is no direct way to inject a test double with this design. We could attempt to get around this, using Java Reflection, but it is best to consider this as TDD feedback on a limitation of our design. The solution is to allow the dependency to be injected, as we have seen in previous examples.

This is often a problem with **legacy code**, which is simply code that has been written before we work on it. If this code has created concrete objects – and the code cannot be changed – then we cannot apply a test double.

Don't test the mock

Testing the mock is a phrase used to describe a test with too many assumptions built into a test double. Suppose we write a stub that stands in for some database access, but that stub contains hundreds of lines of code to emulate detailed specific queries to that database. When we write the test assertions, they will all be based on those detailed queries that we emulated in the stub.

That approach will prove that the SUT logic responds to those queries. But our stub now assumes a great deal about how the real data access code will work. The stub code and the real data access code can quickly get out of step. This results in an invalid unit test that passes but with stubbed responses that can no longer happen in reality.

When to use mock objects

Mocks are useful whenever our SUT is using a push model and requesting an action from some other component, where there is no obvious response such as the following:

- Requesting an action from a remote service, such as sending an email to a mail server
- Inserting or deleting data from a database
- Sending a command over a TCP socket or serial interface
- Invalidating a cache
- Writing logging information either to a log file or distributing logging endpoint

We've learned some techniques in this section that allow us to verify that an action was requested. We have seen how we can use the Dependency Inversion Principle once again to allow us to inject a test double which we can query. We've also seen an example of hand-written code to do this. But must we always write test doubles by hand? In the next section, we will cover a very useful library that does most of the work for us.

Working with Mockito – a popular mocking library

The previous sections have shown examples of using stubs and mocks to test code. We have been writing these test doubles by hand. It's obviously quite repetitive and time-consuming to do this. It begs the question of if this repetitive boilerplate code can be automated away. Thankfully for us, it can. This section will review the help available in the popular Mockito library.

Mockito is a free-of-charge open source library under the MIT license. This license means we can use this for commercial development work, subject to agreement by those we work for. Mockito provides a large range of features aimed at creating test doubles with very little code. The Mockito website can be found at `https://site.mockito.org/`.

Getting started with Mockito

Getting started with Mockito is straightforward. We pull in the `Mockito` library and an extension library in our Gradle file. The extension library allows `Mockito` to integrate closely with *JUnit5*.

The excerpt of `build.gradle` looks like this:

```
dependencies {
    testImplementation 'org.junit.jupiter:junit-jupiter-
api:5.8.2'
    testRuntimeOnly 'org.junit.jupiter:junit-jupiter-
engine:5.8.2'
    testImplementation 'org.assertj:assertj-core:3.22.0'
    testImplementation 'org.mockito:mockito-core:4.8.0'
    testImplementation 'org.mockito:mockito-junit-
jupiter:4.8.0'
}
```

Writing a stub with Mockito

Let's see how Mockito helps us create a stub object. We'll use TDD to create a `UserGreeting` class that delivers a personalized greeting, after fetching our nickname from `interface UserProfiles`.

Let's write this using small steps, to see how TDD and Mockito work together:

1. Write the basic JUnit5 test class and integrate it with Mockito:

    ```
    package examples

    import org.junit.jupiter.api.extension.ExtendWith;
    import org.mockito.junit.jupiter.MockitoExtension;
    ```

```
@ExtendWith(MockitoExtension.class)
public class UserGreetingTest {
}
```

`@ExtendWith(MockitoExtension.class)` marks this test as using Mockito. When we run this JUnit5 test, the annotation ensures that the Mockito library code is run.

2. Add a test confirming the expected behavior. We will capture this in an assertion:

```
package examples;

import org.junit.jupiter.api.Test;
import org.junit.jupiter.api.extension.ExtendWith;
import org.mockito.junit.jupiter.MockitoExtension;
import static org.assertj.core.api.Assertions.assertThat;

@ExtendWith(MockitoExtension.class)
public class UserGreetingTest {

    @Test
    void formatsGreetingWithName() {

        String actual = «»;
        assertThat(actual)
            .isEqualTo("Hello and welcome, Alan");
    }
}
```

This is standard usage of the *JUnit* and *AssertJ* frameworks as we have seen before. If we run the test now, it will fail.

3. Drive out our SUT – the class we want to write – with an Act step:

```
package examples;

import org.junit.jupiter.api.Test;
import org.junit.jupiter.api.extension.ExtendWith;
import org.mockito.junit.jupiter.MockitoExtension;
```

```
import static org.assertj.core.api.Assertions.assertThat;

@ExtendWith(MockitoExtension.class)
public class UserGreetingTest {

    private static final UserId USER_ID
        = new UserId("1234");

    @Test
    void formatsGreetingWithName() {

        var greeting = new UserGreeting();

        String actual =
            greeting.formatGreeting(USER_ID);

        assertThat(actual)
            .isEqualTo("Hello and welcome, Alan");

    }
}
```

This step drives out the two new production code classes, as shown in the following steps.

4. Add a class UserGreeting skeleton:

```
package examples;

public class UserGreeting {
    public String formatGreeting(UserId id) {
        throw new UnsupportedOperationException();
    }
}
```

As usual, we add no code beyond what is required to make our test compile. The design decision captured here shows that our behavior is provided by a formatGreeting() method, which identifies a user by a UserId class.

5. Add a class `UserId` skeleton:

```
package examples;

public class UserId {

    public UserId(String id) {
    }
}
```

Again, we get an empty shell just to get the test to compile. Then, we run the test and it still fails:

Figure 8.3 – Test failure

6. Another design decision to capture is that the `UserGreeting` class will depend on a `UserProfiles` interface. We need to create a field, create the interface skeleton, and inject the field in a new constructor for the SUT:

```
package examples;

import org.junit.jupiter.api.Test;
import org.junit.jupiter.api.extension.ExtendWith;
import org.mockito.junit.jupiter.MockitoExtension;

import static org.assertj.core.api.Assertions.assertThat;

@ExtendWith(MockitoExtension.class)
public class UserGreetingTest {

    private static final UserId USER_ID
        = new UserId("1234");
```

```
private UserProfiles profiles;

@Test
void formatsGreetingWithName() {

    var greeting
        = new UserGreeting(profiles);

    String actual =
        greeting.formatGreeting(USER_ID);

    assertThat(actual)
        .isEqualTo("Hello and welcome, Alan");

    }
}
```

We continue by adding the bare minimum code to get the test to compile. If we run the test, it will still fail. But we've progressed further so the failure is now an UnsupportedOperationException error. This confirms that formatGreeting() has been called:

Figure 8.4 – Failure confirms method call

7. Add behavior to the formatGreeting() method:

```
package examples;

public class UserGreeting {

    private final UserProfiles profiles;

    public UserGreeting(UserProfiles profiles) {
        this.profiles = profiles;
    }
```

```
public String formatGreeting(UserId id) {
    return String.format("Hello and Welcome, %s",
        profiles.fetchNicknameFor(id));
    }
}
```

8. Add `fetchNicknameFor()` to the `UserProfiles` interface:

```
package examples;

public interface UserProfiles {
    String fetchNicknameFor(UserId id);
}
```

9. Run the test. It will fail with a null exception:

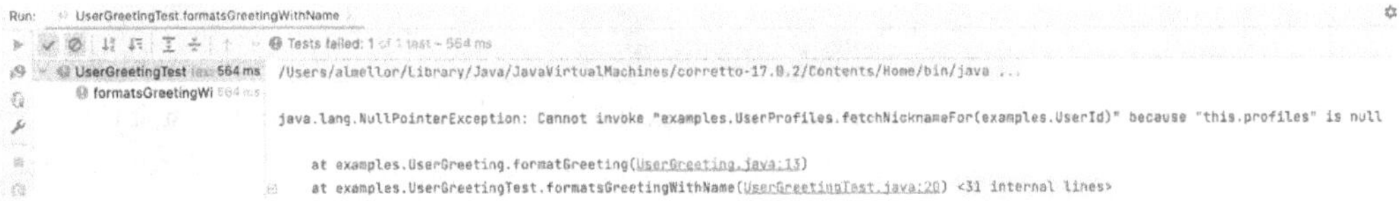

Figure 8.5 – Null exception failure

The test fails because we passed the `profiles` field as a dependency into our SUT, but that field has never been initialized. This is where Mockito comes into play (finally).

10. Add the `@Mock` annotation to the `profiles` field:

```
package examples;

import org.junit.jupiter.api.Test;
import org.junit.jupiter.api.extension.ExtendWith;
import org.mockito.Mock;
import org.mockito.junit.jupiter.MockitoExtension;

import static org.assertj.core.api.Assertions.assertThat;

@ExtendWith(MockitoExtension.class)
public class UserGreetingTest {
```

```
private static final UserId USER_ID = new
UserId("1234");

@Mock
private UserProfiles profiles;

@Test
void formatsGreetingWithName() {

    var greeting = new UserGreeting(profiles);

    String actual =
            greeting.formatGreeting(USER_ID);

    assertThat(actual)
            .isEqualTo("Hello and welcome, Alan");

}
}
```

Running the test now produces a different failure, as we have not yet configured the Mockito mock:

Figure 8.6 – Added mock, not configured

11. Configure @Mock to return the correct stub data for our test:

```
package examples;

import org.junit.jupiter.api.Test;
import org.junit.jupiter.api.extension.ExtendWith;
import org.mockito.Mock;
```

```java
import org.mockito.Mockito;
import org.mockito.junit.jupiter.MockitoExtension;

import static org.assertj.core.api.Assertions.assertThat;
import static org.mockito.Mockito.*;

@ExtendWith(MockitoExtension.class)
public class UserGreetingTest {

    private static final UserId USER_ID = new
    UserId("1234");

    @Mock
    private UserProfiles profiles;

    @Test
    void formatsGreetingWithName() {
        when(profiles.fetchNicknameFor(USER_ID))
            .thenReturn("Alan");


        var greeting = new UserGreeting(profiles);

        String actual =
                greeting.formatGreeting(USER_ID);

        assertThat(actual)
                .isEqualTo("Hello and welcome, Alan");

    }
}
```

12. If you run the test again, it will fail due to a mistake in the greeting text. Fix this and then re-run the test, and it will pass:

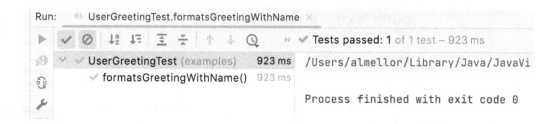

Run: UserGreetingTest.formatsGreetingWithName ✕

✓ ⊘ ↓₂ ↓₌ ⊼ ⊻ ↑ ↓ ⊕ » ✓ Tests passed: 1 of 1 test – 923 ms

✓ UserGreetingTest (examples)   923 ms   /Users/almellor/Library/Java/JavaVi
    ✓ formatsGreetingWithName() 923 ms
                                          Process finished with exit code 0

Figure 8.7 – Test pass

We've just created class UserGreeting, which accesses some stored nicknames for the user, via interface UserProfiles. That interface used DIP to isolate UserGreeting from any implementation details of that store. We used a stub implementation to write the test. We've followed TDD and leveraged Mockito to write that stub for us.

You'll also notice that the test failed in the final step. I expected that step to pass. It didn't because I had typed the greeting message incorrectly. Once again, TDD came to my rescue.

Writing a mock with Mockito

Mockito can create mock objects just as easily as stubs. We can still use the @Mock annotation on a field we wish to become a mock – perhaps making sense of the annotation, at last. We use the Mockito verify() method to check that our SUT called an expected method on a collaborator.

Let's look at how a mock is used. We'll write a test for some SUT code that we expect to send an email via MailServer:

```
@ExtendWith(MockitoExtension.class)
class WelcomeEmailTest {
    @Mock
    private MailServer mailServer;

    @Test
    public void sendsWelcomeEmail() {
        var notifications
                = new UserNotifications( mailServer );

        notifications.welcomeNewUser("test@example.com");

        verify(mailServer).sendEmail("test@example.com",
                "Welcome!",
```

```
                              "Welcome to your account");
        }
    }
```

In this test, we see the @ExtendWith(MockitoExtension.class) annotation to initialize Mockito, and the familiar Arrange, Act and Assert format of our test method. The new idea here is in the assertion. We use the verify() method from the Mockito library to check that the sendEmail() method was called correctly by our SUT. The check also verifies that it was called with the correct parameter values.

Mockito uses code generation to achieve all this. It wraps the interface we labeled with the @Mock annotation and intercepts each and every call. It stores parameter values for each call. When we come to using the verify() method to confirm that the method was called correctly, Mockito has all the data it needs to do this.

Beware Mockito's when() and verify() syntax!

Mockito has subtly different syntax for when() and verify():

* when(**object.method()**).thenReturn(expected value);

* verify(**object**).method();

Blurring the distinction between stubs and mocks

One thing to note about Mockito terminology is that it blurs the distinction between a stub and a mock object. In Mockito, we create test doubles that are labeled as mock objects. But in our test, we can use these doubles as either a stub, a mock, or even a mixture of both.

Setting up a test double to be both a stub and a mock is a test code smell. It's not wrong, but it's worth a pause for thought. We should consider if the collaborator that we are both mocking and stubbing has mixed up some responsibilities. It may be beneficial to split that object up.

Argument matchers – customizing behavior of test doubles

So far, we have configured Mockito test doubles to respond to very specific inputs to the methods they replace. The previous MailServer example checked for three specific parameter values being passed to the sendEmail() method call. But we sometimes want more flexibility in our test doubles.

Mockito provides library methods called argument matchers. These are static methods that are used inside when() and verify() statements. Argument matchers are used to instruct Mockito to respond to a range of parameter values – including nulls and unknown values – that might get passed into a method under test.

The following test uses an argument matcher that accepts any value of `UserId`:

```
package examples2;

import examples.UserGreeting;
import examples.UserId;
import examples.UserProfiles;
import org.junit.jupiter.api.Test;
import org.junit.jupiter.api.extension.ExtendWith;
import org.mockito.Mock;
import org.mockito.junit.jupiter.MockitoExtension;

import static org.assertj.core.api.Assertions.assertThat;
import static org.mockito.ArgumentMatchers.any;
import static org.mockito.Mockito.when;

@ExtendWith(MockitoExtension.class)
public class UserGreetingTest {
    @Mock
    private UserProfiles profiles;

    @Test
    void formatsGreetingWithName() {
      when(profiles.fetchNicknameFor(any()))
          .thenReturn("Alan");

        var greeting = new UserGreeting(profiles);

        String actual =
          greeting.formatGreeting(new UserId(""));

        assertThat(actual)
          .isEqualTo("Hello and welcome, Alan");
    }
}
```

We've added an any() argument matcher to the stubbing of the fetchNicknameFor() method. This instructs Mockito to return the expected value Alan no matter what parameter value is passed into fetchNicknameFor(). This is useful when writing tests to guide our readers and help them to understand what is important and what is not for a particular test.

Mockito offers a number of argument matchers, described in the Mockito official documentation. These argument matchers are especially useful when creating a stub to simulate an error condition. This is the subject of the next section.

Driving error handling code with tests

In this section, we're going to look into a great use of stub objects, which is their role in testing error conditions.

As we create our code, we need to ensure that it handles error conditions well. Some error conditions are easy to test. An example might be a user input validator. To test that it handles the error caused by invalid data, we simply write a test that feeds it invalid data and then write an assertion to check it successfully reported the data was invalid. But what about the code that uses it?

If our SUT is code that responds to an error condition raised by one of its collaborators, we need to test that error response. How we test it depends on the mechanism we chose to report that error. We may be using a simple status code, in which case returning that error code from a stub will work very well.

We may also have chosen to use Java exceptions to report this error. Exceptions are controversial. If misused, they can lead to very unclear control flow in your code. We need to know how to test them, however, as they appear in several Java libraries and in-house coding styles. Fortunately, there's nothing difficult about writing the test for exception-handling code.

We start with creating a stub, using any of the approaches covered in this chapter. We then need to arrange for the stub to throw the appropriate exception when we call a method. Mockito has a nice feature to do this, so let's see an example Mockito test that uses exceptions:

```
@Test
public void rejectsInvalidEmailRecipient() {
    doThrow(new IllegalArgumentException())
        .when(mailServer).sendEmail(any(),any(),any());


    var notifications
        = new UserNotifications( mailServer );


    assertThatExceptionOfType(NotificationFailureException.
class)
            .isThrownBy(()->notifications
```

```
                    .welcomeNewUser("not-an-email-address"));
}
```

At the start of this test, we use Mockito doThrow() to configure our mock object. This configures the Mockito mock object mailServer to throw IllegalArgumentException whenever we call sendEmail(), no matter what parameter values we send. This reflects a design decision to make sendEmail() throw that exception as a mechanism to report that the email address was not valid. When our SUT calls mailServer.sendEmail(), that method will throw IllegalArgumentExeption. We can exercise the code that handles this.

For this example, we decided to make the SUT wrap and rethrow IllegalArgumentException. We choose to create a new exception that relates to the responsibility of user notifications. We will call it NotificationFailureException. The assertion step of the test then uses the AssertJ library feature assertThatExceptionOfType(). This performs the Act and Assert steps together. We call our SUT welcomeNewUser() method and assert that it throws our NotificationFailureException error.

We can see how this is enough to trigger the exception-handling response in our SUT code. This means we can write our test and then drive out the required code. The code we write will include a catch handler for InvalidArgumentException. In this case, all the new code has to do is throw a NotificationFailureException error. This is a new class that we will create that extends RuntimeException. We do this to report that something went wrong by sending a notification. As part of normal system layering considerations, we want to replace the original exception with a more general one, which is better suited to this layer of code.

This section has examined features of Mockito and AssertJ libraries that help us use TDD to drive out exception-handling behavior. In the next section, let's apply this to an error in our Wordz application.

Testing an error condition in Wordz

In this section, we will apply what we've learned by writing a test for a class that will choose a random word for the player to guess, from a stored set of words. We will create an interface called WordRepository to access stored words. We will do this through a fetchWordByNumber(wordNumber) method, where wordNumber identifies a word. The design decision here is that every word is stored with a sequential number starting from 1 to help us pick one at random.

We will be writing a WordSelection class, which is responsible for picking a random number and using that to fetch a word from storage that is tagged with that number. We will be using our RandomNumbers interface from earlier. For this example, our test will cover the case where we attempt to fetch a word from the WordRepository interface, but for some reason, it isn't there.

We can write the test as follows:

```java
@ExtendWith(MockitoExtension.class)
public class WordSelectionTest {

    @Mock
    private WordRepository repository;

    @Mock
    private RandomNumbers random;

    @Test
    public void reportsWordNotFound() {
        doThrow(new WordRepositoryException())
                .when(repository)
                .fetchWordByNumber(anyInt());

        var selection = new WordSelection(repository,
                                          random);

        assertThatExceptionOfType(WordSelectionException.class)
                .isThrownBy(
                        () ->selection.getRandomWord());
    }
}
```

The test captures a few more design decisions relating to how we intend WordRepository and WordSelection to work. Our fetchWordByNumber(wordNumber) repository method will throw WordRepositoryException if there are any problems retrieving the word. Our intention is to make WordSelection throw its own custom exception to report that it cannot complete the getRandomWord() request.

To set this situation up in the test, we first arrange for the repository to throw. This is done using the Mockito doThrow() feature. Whenever the fetchWordByNumber() method is called, whatever parameter we pass into it Mockito will throw the exception we asked it to throw, which is WordRepositoryException. This allows us to drive out the code that handles this error condition.

Our Arrange step is completed by creating the `WordSelection` SUT class. We pass in two collaborators to the constructor: the `WordRepository` instance and a `RandomNumbers` instance. We have asked Mockito to create stubs for both interfaces by adding the `@Mock` annotation to test double the `repository` and `random` fields.

With the SUT now properly constructed, we are ready to write the Act and Assert steps of the test. We are testing that an exception is thrown, so we need to use the `assertThatExceptionOfType()` AssertJ facility to do this. We can pass in the class of the exception that we are expecting to be thrown, which is `WordSelectionException`. We chain the `isThrownBy()` method to perform the Act step and make our SUT code run. This is provided as a Java lambda function as a parameter to the `isThrownBy()` method. This will call the `getRandomWord()` method, which we intend to fail and throw an exception. The assertion will confirm that this has happened and that the expected kind of exception class has been thrown. We will run the test, see it fail, and then add the necessary logic to make the test pass.

The test code shows us that we can use test doubles and verification of error conditions with test-first TDD. It also shows that tests can become easily coupled to a specific implementation of a solution. There are a lot of design decisions in this test about which exceptions happen and where they are used. These decisions even include the fact that exceptions are being used at all to report errors. All that said, this is still a reasonable way to split responsibilities and define contracts between components. It is all captured in the test.

Summary

In this chapter, we've looked at how to solve the problem of testing problematic collaborators. We have learned how to use stand-in objects for collaborators called test doubles. We've learned that this gives us simple control over what those collaborators do inside our test code.

Two kinds of test double are especially useful to us: the stub and the mock. Stubs return data. Mocks verify that methods were called. We've learned how to use the Mockito library to create stubs and mocks for us.

We've used AssertJ to verify the SUT behaved correctly under the various conditions of our test doubles. We've learned how to test error conditions that throw exceptions.

These techniques have expanded our toolkit for writing tests.

In the next chapter, we are going to cover a very useful system design technique that allows us to get most of our code under FIRST unit test, and at the same time avoid the problems of testing collaborations with external systems that we cannot control.

Questions and answers

1. Are the terms stub and mock used interchangeably?

    Yes, even though they have different meanings. In normal conversation, we tend to trade precision for fluency, and that's okay. It's important to understand the different uses that each kind of test double has. When speaking, it's usually better to not be pedantic whenever a group of people knows what is meant. So long as we stay aware that a test double is the proper general term and that the specific types of doubles have different roles, all will be well.

2. What is the problem known as "testing the mock"?

    This happens when the SUT has no real logic in it, yet we try to write a unit test anyway. We wire up a test double to the SUT and write the test. What we will find is that the assertions only check that the test double-returned the right data. It's an indication that we have tested at the wrong level. This kind of error can be driven by setting unwise code coverage targets or forcing an equally unwise test-per-method rule. This test adds no value and should be removed.

3. Can test doubles be used anywhere?

    No. This only works if you have designed your code using the Dependency Inversion Principle so that a test double can be swapped in place of a production object. Using TDD certainly forces us to think about this kind of design issue early.

    Writing tests later is made more difficult if there is insufficient access to inject test doubles where they are needed. Legacy code is particularly difficult in this respect, and I recommend reading the book *Working Effectively with Legacy Code* by Michael Feathers for techniques to aid in adding tests to code that lacks the necessary test access points. (See the *Further reading* list.)

Further reading

*   https://site.mockito.org/

    Mockito library home page

*   *Working Effectively with Legacy Code, Michael C. Feathers ISBN 978-0131177055*

    This book explains how you can work with legacy code written without Dependency Inversion access points for test doubles. It shows a range of techniques to safely rework the legacy code so that test doubles can be introduced.

9

Hexagonal Architecture – Decoupling External Systems

We've already learned how to write tests using the arrange, act, and assert template. We've also learned about some software design principles, known as the SOLID principles, that help us break our software down into smaller components. Finally, we've learned how test doubles can stand in for collaborating components to make FIRST unit tests easier to write. In this chapter, we're going to combine all those techniques into a powerful design approach known as the hexagonal architecture.

Using this approach, we will benefit from getting more of our application logic under unit tests and reducing the number of integration and end-to-end tests required. We will build in a natural resilience to changes outside our application. Development chores such as changing a database supplier will be simplified, by having fewer places where our code needs to be changed. We will also be able to unit test across larger units, bringing some tests that require end-to-end testing in other approaches under unit tests instead.

In this chapter, we're going to cover the following main topics:

- Why external systems are difficult
- Dependency inversion to the rescue
- Abstracting out the external system
- Writing the domain code
- Substituting test doubles for external systems
- Unit testing bigger units
- Wordz – abstracting the database

Technical requirements

The code for this chapter can be found at `https://github.com/PacktPublishing/Test-Driven-Development-with-Java/tree/main/chapter09`.

Why external systems are difficult

In this section, we're going to review the driving force behind the hexagonal architecture approach – the difficulty of working with external systems. Dependencies on external systems cause problems in development. The solution leads to a nice design approach.

Let's look at a simple way of handling external systems. The task of our user is to pull a report of this month's sales from a database. We will write one piece of code that does exactly that. The software design looks like this:

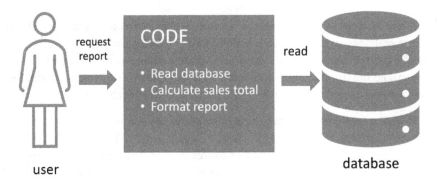

Figure 9.1 – One piece of code does everything

In this design, we have sales data stored in a database in the usual way. We write some code to pull the report on behalf of our user. It is a single piece of code that does the whole job as a single step. It will connect to the database, send a query, receive the results, do some processing, and format the results ready for the user to read.

On the plus side, we know this style of coding works. It will achieve its aim of providing that sales report to the user. On the downside, the code combines three different responsibilities – accessing a database, performing logic, and formatting a report. It might mix up SQL statements to the database with `html5` tags to make a formatted report. As we saw in a previous chapter, this can make future code changes in one area ripple out and impact the other areas. Ideally, that should not happen. But the real challenge is writing a test for this one piece of code. We'll need to parse and understand whatever format we send the report to the user in. We'll also need to work directly with that database.

In the following subsections, we'll review some wider challenges that external systems present to testing. These include environmental problems, accidental transactions, uncertain data, operating system calls, and third-party libraries.

Environmental problems bring trouble

The environment that our software runs in often causes challenges. Suppose our code reads data from a database. Even if the code is correct, it might not be able to read that data, due to problems in the environment beyond our control. Such problems include the following:

- **Network connection dropped**: Many reasons can cause this. Locally, a network cable is pulled out by mistake. Maybe the database is hosted over the internet somewhere, and our ISP has dropped the connection.

- **Power failures**: A power failure on the database server, or a local network switch is enough to put the database out of our reach.

- **Equipment limits**: Maybe the database server itself has run out of disk space and cannot operate. Maybe the exact query we have written is hitting the database in a way that takes a long time to complete, perhaps due to missing indices.

Whatever the cause, if our code cannot access the data in the database, it's not going to work. As this is a possibility, writing a test for our report generation code is made much harder.

Even when our code can access the data in the database, it's not that easy to work with in testing. Suppose we write a test that verifies that we can read the production database correctly, by reading a username. What username would we expect to read? We don't know, because the test is not in control of what data gets added. The available usernames will be whatever names were added by real users. We could make the test add a known test username to the database – but then, we have just created a fake user that real users can interact with. This is not what we want at all.

A database stores data, causing further problems for our tests. Suppose we write a test against a test database, which begins by writing a test username. If we have run this test before, the test username will already be stored in the database. Typically, the database will report a duplicate item error and the test will fail.

Tests against databases need cleaning up. Any test data stored must be deleted after the tests have been completed. If we attempt to delete data after the test has succeeded, the deletion code may never run if the test fails. We could avoid this by always deleting the data before the test runs. Such tests will be slow to run.

Accidentally triggering real transactions from tests

When our code is limited to only accessing a production system, then every time we use that code, something will happen in production. The payment processor may issue charges. Real bank accounts may become debited. Alarms may be activated, causing real evacuations. In a famous example from Hawaii, a system test triggered a real text message saying Hawaii was under missile attack – which it wasn't. This is serious stuff.

> **Hawaii false missile attack warning**
>
> For details on this example of testing going wrong, see `https://en.wikipedia.org/wiki/2018_Hawaii_false_missile_alert`.

Accidental real transactions can result in real losses to a company. They could end up as losses to the 3Rs of a business – revenue, reputation, and retention. None of those are good. Our tests mustn't accidentally trigger real consequences from production systems.

What data should we expect?

In our sales report example, the biggest problem with writing a test is that we would need to know what the correct answer is to the monthly sales report in advance. How do we do that when we are connected to the production system? The answer will be whatever the sales report says it is. We have no other way of knowing.

The fact that we need the sales report code to be working correctly before we can test that the sales report code is working correctly is a big problem here! This is a circular dependency we cannot break.

Operating system calls and system time

Sometimes, our code may need to make calls to the operating system to do its job. Perhaps it needs to delete all the files in a directory from time to time or it may be dependent on the system time. An example would be a log file cleanup utility, which runs every Monday at 02:00 A.M. The utility will delete every file in the `/logfiles/` directory.

Testing such a utility would be difficult. We would have to wait until 02:00 A.M. on Monday and verify that all the log files have been deleted. While we could make this work, it isn't very effective. It would be nice to find a better approach that allowed us to test anytime we liked, ideally without deleting any files.

Challenges with third-party services

A common task in business software is to accept payment from a customer. For that, we inevitably use a third-party payment processor such as PayPal or Stripe, as two examples. In addition to the challenges of network connectivity, third-party APIs provide us with further challenges:

- **Service downtime**: Many third-party APIs will have a period of scheduled maintenance where the service is unavailable for a time. That spells "test failed" for us.

- **API changes**: Suppose our code uses API version 1 and API version 2 is pushed live. Our code will still be using version 1 calls, which might no longer work on version 2 of the API. Now, that is considered rather bad practice – it's called breaking a published interface – but it can and does happen. Worse, with our one piece of code, the version 2 changes might cause changes everywhere in our code.

- **Slow responses**: If our code makes an API call to an external service, there is always a possibility that the response will come back later than expected by our code. Our code will fail in some way usually and cause tests to fail.

Plenty of challenges exist when we mix external services and a single monolithic piece of code, complicating both maintenance and testing. The question is what can we do about it? The next section looks at how the **Dependency Inversion Principle** can help us follow a design approach known as a hexagonal architecture, which makes external systems easier to deal with.

Dependency inversion to the rescue

In this section, we will review a design approach known as the hexagonal architecture, based on the SOLID principles we already know. Using this approach allows us to use TDD more effectively across more of our code base.

We learned about the **Dependency Inversion Principle** previously in this book. We saw that it helps us isolate some code we wanted to test from the details of its collaborators. We noted that was useful for testing things that connected to external systems that were outside of our control. We saw how the single responsibility principle guided us into splitting up software into smaller, more focused tasks.

Applying these ideas to our earlier sales reporting example, we would arrive at an improved design, as shown in the following diagram:

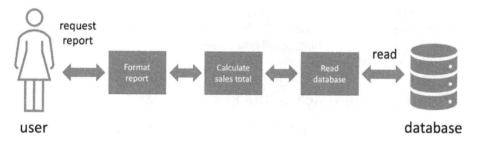

Figure 9.2 – Applying SOLID to our sales report

The preceding diagram shows how we have applied SOLID principles to splitting up our sales report code. We have used the single responsibility principle to break down the overall task into three separate tasks:

- Formatting the report
- Calculating the sales total
- Reading the sales data from the database

This already makes the application a little easier to work with. More importantly, we've isolated the code that calculates the sales total from both the user and the database. This calculation no longer directly accesses the database. It goes through another piece of code responsible for doing only that. Likewise, the calculation result isn't directly formatted and sent to the user. Another piece of code is responsible for that.

We can apply the **Dependency Inversion Principle** here as well. By inverting the dependencies on the formatting and database access code, our calculated sales total is now free from knowing any of their details. We've made a significant breakthrough:

- The calculation code is now fully isolated from the database and formatting
- We can swap in any piece of code that can access any database
- We can swap in any piece of code that can format a report
- We can use test doubles in place of the formatting and database access code

The biggest benefit is that we can swap in any piece of code that can access any database, without changing the calculation code. For example, we could change from a Postgres SQL database to a Mongo NoSQL database without changing the calculation code. We can use a test double for the database so that we can test the calculation code as a FIRST unit test. These are very significant advantages, not just in terms of TDD and testing, but also in terms of how our code is organized. Considering the one-piece sales report solution to this one, we have moved from pure writing code to software engineering. We're thinking beyond just getting code to work and focusing on making code easy to work with. The next few subsections will look at how we can generalize this approach, resulting in the hexagonal architecture. We will understand how this approach delivers a logical organization of code that helps us apply TDD more effectively.

Generalizing this approach to the hexagonal architecture

This combination of the single responsibility principle and dependency inversion seems to have brought us some benefits. Could we extend this approach to the entire application and get the same benefits? Could we find a way to separate all our application logic and data representations from the constraints of external influence? We most certainly can, and the general form of this design is shown in the following diagram:

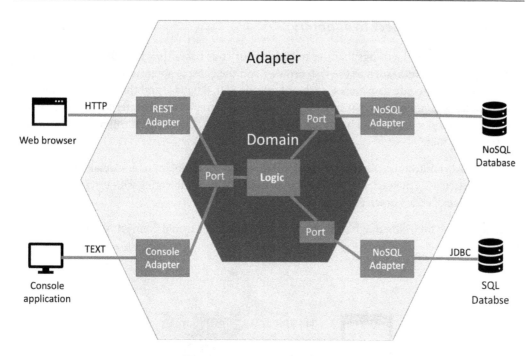

Figure 9.3 – Hexagonal architecture

The preceding diagram shows what happens when we generalize the use of dependency inversion and single responsibility to an entire application. It is called the hexagonal architecture, also known as ports and adapters after the original term used by Alastair Cockburn, who first described this approach. The benefit is that it completely isolates the core logic of our application from the details of external systems. This helps us with testing that core logic. It also provides a reasonable template for a well-engineered design for our code.

Overview of the hexagonal architecture's components

To provide us with this isolation of our core application logic, the hexagonal architecture divides the whole program into four spaces:

- External systems, including web browsers, databases, and other computing services
- Adapters implement the specific APIs required by the external systems
- Ports are the abstraction of what our application needs from the external system
- The domain model contains our application logic, free of external system details

The central core of our application is the domain model, surrounded by the support it needs from external systems. It indirectly uses but is not defined by those external systems. Let's walk through each component in the hexagonal architecture in more detail, to understand what each one is and is not responsible for.

External systems connect to adapters

External systems are all the things that live outside of our code base. They include things that the user directly interacts with, such as the web browser and the console application in the preceding diagram. They also include data stores, such as both the SQL database and the NoSQL database. Other examples of common external systems include desktop graphical user interfaces, filesystems, downstream web service APIs, and hardware device drivers. Most applications will need to interact with systems like these.

In the hexagonal architecture, the core of our application code does not know any details about how the external systems are interacted with. The responsibility of communicating with external systems is given to a piece of code known as an adapter.

As an example, the following diagram shows how a web browser would connect to our code via a REST adapter:

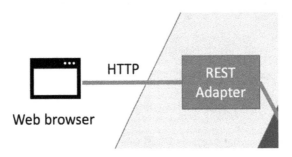

Figure 9.4 – Browser connecting to a REST adapter

In the preceding diagram, we can see the web browser connecting to a REST adapter. This adapter understands HTTP requests and responses, which are the very core of the web. It also understands the JSON data format, often using libraries to convert the JSON data into some internal representation for our code. This adapter will also understand the specific protocol that we will have designed for our application's REST API – the precise sequence of HTTP verbs, responses, status codes, and JSON-encoded payload data we come up with as an API.

> **Note**
> Adapters encapsulate all the knowledge our system needs to interact with an external system – and nothing else. This knowledge is defined by the external system's specifications. Some of those may be designed by ourselves.

Adapters have the single responsibility of knowing how to interact with an external system. If that external system changes its public interface, only our adapter will need to change.

Adapters connect to ports

Moving toward the domain model, adapters connect to ports. Ports are part of the domain model. They abstract away the details of the adapter's intricate knowledge of its external system. Ports answer a slightly different question: what do we need that external system for? The ports use the **Dependency Inversion Principle** to isolate our domain code from knowing any details about the adapters. They are written purely in terms of our domain model:

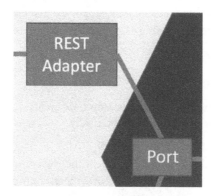

Figure 9.5 – Adapters connect to ports

The REST adapter described previously encapsulates the details of running a REST API, using knowledge of HTTP and JSON. It connects to a commands port, which provides our abstraction of commands coming in from the web – or anywhere else, for that matter. Given our sales report example earlier, the commands port would include a technology-free way of requesting a sales report. In code, it might look as simple as this:

```
package com.sales.domain;

import java.time.LocalDate;

public interface Commands {
    SalesReport calculateForPeriod(LocalDate start,
                                   LocalDate end);
}
```

This code fragment features the following:

- No references to HttpServletRequest or anything to do with HTTP
- No references to JSON formats

- References to our domain model – `SalesReport` and `java.time.LocalDate`
- The `public` access modifier, so it can be called from the REST adapter

This interface is a port. It gives us a general-purpose way to get a sales report from our application. Referring to *Figure 9.3*, we can see that the console adapter also connects to this port, providing the user with a command-line interface to our application. The reason is that while users can access our application using different kinds of external systems – the web and the command line – our application does the same thing in either case. It only supports one set of commands, no matter where those commands are requested from. Fetching a `SalesReport` object is just that, no matter which technology you request it from.

> **Note**
>
> Ports provide a logical view of what our application needs from an external system, without constraining how those needs should be met technically.

Ports are where we invert dependencies. Ports represent the reason our domain model needs those external systems. If the adapters represent the how, ports represent the why.

Ports connect to our domain model

The final step in the chain is connecting to the domain model itself. This is where our application logic lives. Think of it as pure logic for the problem our application is solving. Because of the ports and adapters, the domain logic is unconstrained by details of external systems:

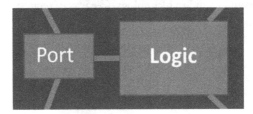

Figure 9.6 – Ports connect to the domain model

The domain model represents the things our users want to do, in code. Every user story is described by code here. Ideally, the code in this layer uses the language of the problem we are solving, instead of technology details. When we do this well, this code becomes storytelling – it describes actions our users care about in terms they have told us about. It uses their language – the language of our users – not obscure computer language.

The domain model can contain code written in any paradigm. It might use **functional programming** (**FP**) ideas. It may even use **object-oriented programming** (**OOP**) ideas. It might be procedural. It might even use an off-the-shelf library that we configure declaratively. My current style is to use OOP for the overall structure and organization of a program, then use FP ideas inside the object methods to implement them. It makes no difference to either the hexagonal architecture or TDD how we implement this domain model. Whatever way suits your coding style is just fine here, so long as you use the ideas of ports and adapters.

> **Note**
> The domain model contains code that describes how the user's problem is being solved. This is the essential logic of our application that creates business value.

At the center of the entire application is the domain model. It contains the logic that brings the user's stories to life.

The golden rule – the domain never connects directly to adapters

To preserve the benefits of isolating the domain model from adapters and external systems, we follow one simple rule: the domain model never connects directly to any of the adapters. This is always done through a port.

When our code follows this design approach, it is straightforward to check whether we've got the ports and adapters split right. We can make two high-level structural decisions:

- The domain model lives in a domain package (and sub packages)
- The adapters live in an adapters package (and sub packages)

We can analyze the code to check that anything in the domain package contains no import statements from the adapters package. Import checks can be done visually in code reviews or pairing/mobbing. Static analysis tools such as SonarQube can automate import checks as part of the build pipeline.

> **The golden rules of the hexagonal architecture**
> The domain model never connects directly to anything in the adapter layer so that our application logic does not depend on details of external systems.
>
> The adapters connect to ports so that code connecting to external systems is isolated.
>
> Ports are part of the domain model to create abstractions of external systems.
>
> The domain model and the adapters depend on the ports only. This is dependency inversion at work.

These simple rules keep our design in line and preserve the isolation of the domain model.

Why the hexagon shape?

The idea behind the hexagon shape used in the diagram is that each face represents one external system. In terms of a graphical representation of a design, having up to six external systems represented is usually sufficient. The idea of the inner and outer hexagons to represent the domain model and adapter layer shows graphically how the domain model is the core of our application and that it is isolated from external systems by the ports and adapter layer.

The critical idea behind the hexagonal architecture is the ports and adapters technique. The actual number of sides depends on how many external systems there are. The number of those is not important.

In this section, we introduced the hexagonal architecture and the benefits it provides, and provided a general overview of how all the essential pieces fit together. Let's turn to the next section and look specifically at the decisions we need to make to abstract out an external system.

Abstracting out the external system

In this section, we will consider some of the decisions we need to make when applying the hexagonal architecture approach. We'll take a step-by-step approach to handling external systems, where we will first decide what the domain model needs, then work out the right abstractions that hide their technical details. We will consider two common external systems: web requests and database access.

Deciding what our domain model needs

The place to begin our design is with our domain model. We need to devise a suitable port for our domain model to interact with. This port has to be free from any details of our external system, and at the same time, it must answer the question of what our application needs this system for. We are creating an abstraction.

A good way to think about abstractions is to think about what would stay the same if we changed how we performed a task. Suppose we want to eat warm soup for lunch. We might warm it in a pan on the stove or perhaps warm it in the microwave. No matter how we choose to do it, what we are doing stays the same. We are warming the soup and that is the abstraction we're looking for.

We don't often warm soup in software systems unless we are building an automated soup vending machine. But there are several common kinds of abstractions we will be using. This is because common kinds of external systems are used when building a typical web application. The first and most obvious is the connection to the web itself. In most applications, we will encounter some kind of data store, typically a third-party database system. For many applications, we will also be calling out to another web service. In turn, this service may call others in a fleet of services, all internal to our company. Another typical web service call is to a third-party web service provider, such as a credit card payment processor, as an example.

Let's look at ways of abstracting these common external systems.

Abstracting web requests and responses

Our application will respond to HTTP requests and responses. The port we need to design represents the request and the response in terms of our domain model, stripping away the web technology.

Our sales report example could introduce these ideas as two simple domain objects. These requests can be represented by a `RequestSalesReport` class:

```java
package com.sales.domain;

import java.time.LocalDate;

public class RequestSalesReport {
    private final LocalDate start;
    private final LocalDate end;

    public RequestSalesReport(LocalDate start,
                              LocalDate end) {
        this.start = start;
        this.end = end;
    }

    public SalesReport produce(SalesReporting reporting) {
        return reporting.reportForPeriod(start, end);
    }
}
```

Here, we can see the critical pieces of our domain model of the request:

- What we are requesting – that is, a sales report, captured in the class name
- The parameters of that request – that is, the start and end dates of the reporting period

We can see how the response is represented:

- The `SalesReport` class will contain the raw information requested

We can also see what is not present:

- The data formats used in the web request
- HTTP status codes, such as 200 OK
- `HTTPServletRequest` and `HttpServletResponse` or equivalent framework objects

This is a pure domain model representation of a request for a sales report between two dates. There is no hint of this having come from the web, a fact that is very useful as we can request it from other input sources, such as a desktop GUI or a command line. Even better, we can create these domain model objects very easily in a unit test.

The preceding example shows an object-oriented, tell-don't-ask approach. We could just as easily choose an FP approach. If we did, we would represent the request and response as pure data structures. The record facility that was added to Java 17 is well suited to representing such data structures. What's important is that the request and response are written purely in domain model terms – nothing of the web technology should be present.

Abstracting the database

Without data, most applications aren't particularly useful. Without data storage, they become rather forgetful of the data we supply. Accessing data stores such as relational databases and NoSQL databases is a common task in web application development.

In a hexagonal architecture, we start by designing the port that the domain model will interact with, again in pure domain terms. The way to create a database abstraction is to think about what data needs storing and not how it will be stored.

A database port has two components:

- An interface to invert the dependency on the database.

  The interface is often known as a repository. It has also been termed a data access object. Whatever the name, it has the job of isolating the domain model from any part of our database and its access technology.

- Value objects representing the data itself, in domain model terms.

  A value object exists to transfer data from place to place. Two value objects that each hold the same data values are considered equal. They are ideal for transferring data between the database and our code.

Returning to our sales report example, one possible design for our repository would be this:

```
package com.sales.domain;

public interface SalesRepository {
    List<Sale> allWithinDateRange(LocalDate start,
                                  LocalDate end);
}
```

Here, we have a method called `allWithinDateRange()` that allows us to fetch a set of individual sales transactions falling within a particular date range. The data is returned as `java.util.List` of simple `Sale` value objects. These are fully featured domain model objects. They may well have methods on them that perform some of the critical application logic. They may be little more than basic data structures, perhaps using a Java 17 `record` structure. This choice is part of our job in deciding what a well-engineered design looks like in our specific case.

Again, we can see what is not present:

- Database connection strings

- JDBC or JPA API details – the standard Java Database Connectivity library

- SQL queries (or NoSQL queries)

- Database schema and table names

- Database stored procedure details

Our repository designs focus on what our domain model needs our database to provide but does not constrain how it provides. As a result, some interesting decisions have to be taken in designing our repository, concerning how much work we put into the database and how much we do in the domain model itself. Examples of this include deciding whether we will write a complex query in the database adapter, or whether we will write simpler ones and perform additional work in the domain model. Likewise, will we make use of stored procedures in the database?

Whatever trade-offs we decide in these decisions, once again, the database adapter is where all those decisions reside. The adapter is where we see the database connection strings, query strings, table names, and so on. The adapter encapsulates the design details of our data schema and database technology.

Abstracting calls to web services

Making calls to other web services is a frequent development task. Examples include calls to payment processors and address lookup services. Sometimes, these are third-party external services, and sometimes, they live inside our web service fleet. Either way, they generally require some HTTP calls to be made from our application.

Abstracting these calls proceeds along similar lines to abstracting the database. Our port is made up of an interface that inverts the dependency on the web service we are calling, and some value objects that transfer data.

An example of abstracting a call to a mapping API such as Google Maps, for example, might look like this:

```
package com.sales.domain;


public interface MappingService {
```

```
    void addReview(GeographicLocation location,
            Review review);
}
```

We have an interface representing `MappingService` as a whole. We've added a method to add a review of a particular location on whichever service provider we end up using. We're using `GeographicLocation` to represent a place, defined in our terms. It may well have a latitude and longitude pair in it or it may be based on postal code. That's another design decision. Again, we see no sign of the underlying map service or its API details. That code lives in the adapter, which would connect to the real external mapping web service.

This abstraction offers us benefits in being able to use a test double for that external service and being able to change service providers in the future. You never know when an external service might shut down or become too costly to use. It's nice to keep our options open by using the hexagonal architecture.

This section has presented some ideas for the most common tasks in working with external systems in a hexagonal architecture. In the next section, we'll discuss general approaches to writing code in the domain model.

Writing the domain code

In this section, we will look at some of the things we need to think about as we write the code for our domain model. We'll cover what kinds of libraries we should and should not use in the domain model, how we deal with application configuration and initialization, and we'll also think about what impact popular frameworks have.

Deciding what should be in our domain model

Our domain model is the very core of our application and the hexagonal architecture puts it up front and center. A good domain model is written using the language of our users' problem domain; that's where the name comes from. We should see the names of program elements that our users would recognize. We should recognize the problem being solved over and above the mechanisms we are using to solve it. Ideally, we will see terms from our user stories being used in our domain model.

Applying the hexagonal architecture, we choose our domain model to be independent of those things that are not essential to solving the problem. That's why external systems are isolated. We may initially think that creating a sales report means that we must read a file and we must create an HTML document. But that's not the essential heart of the problem. We simply need to get sales data from somewhere, perform some calculations to get totals for our report, then format it somehow. The somewhere and somehow can change, without affecting the essence of our solution.

Bearing this constraint in mind, we can take any standard analysis and design approach. We are free to choose objects or decompose them into functions as we normally do. We only have to preserve that distinction between the essence of the problem and the implementation details.

We need to exercise judgment in these decisions. In our sales report example, the source of the sales data is of no consequence. As a counter-example, suppose we are making a linter for our Java program files – it's quite reasonable to have the concept of files represented directly in our domain model. This problem domain is all about working with Java files, so we should make that clear. We may still decouple the domain model of a file from the OS-specific details of reading and writing it, but the concept would be in the domain model.

Using libraries and frameworks in the domain model

The domain model can use any pre-written library or framework to help do its job. Popular libraries such as Apache Commons or the Java Standard Runtime library generally present no problems here. However, we need to be aware of frameworks that bind us to the world of external systems and our adapter layer. We need to invert dependencies on those frameworks, leaving them to be just an implementation detail of the adapter layer.

An example might be the @RestController annotation of Spring Boot. It looks like pure domain code at first sight, but it ties the class tightly to generated code that is specific to the web adapter.

Deciding on a programming approach

The domain model can be written using any programming paradigm. This flexibility means that we will need to decide on which approach to use. This is never a purely technical decision, like with so many things in software. We should consider the following:

- **Existing team skills and preferences**: What paradigm does the team know best? Which paradigm would they like to use, given the chance?

- **Existing libraries, frameworks, and code bases**: If we are going to be using pre-written code – and let's face it, we almost certainly will – then what paradigm would best suit that code?

- **Style guides and other code mandates**: Are we working with an existing style guide or paradigm? If we are being paid for our work – or we are contributing to an existing open source project – we will need to adopt the paradigm set out for us.

The good news is that whatever paradigm we choose, we will be able to write our domain model successfully. While the code may look different, equivalent functionality can be written using any of the paradigms.

Substituting test doubles for external systems

In this section, we'll discuss one of the biggest advantages that the hexagonal architecture brings to TDD: high testability. It also brings some workflow advantages.

Replacing the adapters with test doubles

The key advantage the hexagonal architecture brings to TDD is that it is trivially easy to replace all the adapters with test doubles, giving us the ability to test the entire domain model with FIRST unit tests. We can test the entire application core logic without test environments, test databases, or HTTP tools such as Postman or curl – just fast, repeatable unit tests. Our testing setup looks like this:

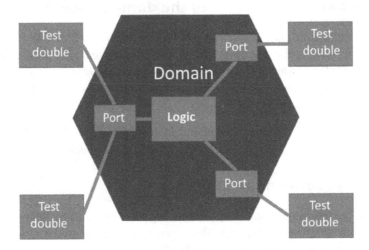

Figure 9.7 – Testing the domain model

We can see that all the adapters have been replaced by test doubles, completely freeing us from our environment of external systems. Unit tests can now cover the whole domain model, reducing the need for integration tests.

We gain several benefits by doing this:

- **We can write TDD tests first with ease**: There's no friction in writing a simple test double that lives entirely in memory and has no dependencies on the test environment.

- **We gain FIRST unit test benefits**: Our tests run very fast indeed and are repeatable. Typically, testing an entire domain model takes the order of seconds, not hours. The tests will repeatably pass or fail, meaning we are never wondering whether a build failure was due to a flaky integration test failure.

- **It unlocks our team**: We can do useful work building the core logic of our system, without having to wait for test environments to be designed and built.

The techniques for creating the test doubles were outlined in *Chapter 8, Test Doubles – Stubs and Mocks*. There is nothing new required in terms of implementing these doubles.

One consequence of being able to test the whole domain model is that we can apply TDD and FIRST unit tests to much larger program units. The next section discusses what that means for us.

Unit testing bigger units

The previous section introduced the idea of surrounding our domain model with test doubles for every port. This gives us some interesting opportunities to discuss in this section. We can test units that are as large as a user story.

We're familiar with unit tests as being things that test in the small. There's a good chance you'll have heard somebody say that a unit test should only ever apply to a single function, or that every class should have one unit test for every method. We've already seen how that's not the best way to use unit tests. Tests like those miss out on some advantages. We are better served by thinking of tests as covering behavior instead.

The combined approach of designing with the hexagonal architecture and testing behaviors instead of implementation details leads to an interesting system layering. Instead of having traditional layers, as we might do in a three-tier architecture, we have circles of increasingly higher-level behavior. Inside our domain model, we will find those tests-in-the-small. But as we move outward, toward the adapter layer, we will find bigger units of behavior.

Unit testing entire user stories

The ports in the domain model form a natural high-level boundary of the domain model. If we review what we've learned in this chapter, we'll see that this boundary consists of the following:

- The essence of requests from users
- The essence of a response from our application
- The essence of how data needs storing and accessing
- All using technology-free code

This layer is the essence of what our application does, free from the details of how it does it. It is nothing less than the original user stories themselves. The most significant thing about this domain model is that we can write FIRST unit tests against it. We have all we need to replace difficult-to-test external systems with simple test doubles. We can write unit tests that cover entire user stories, confirming that our core logic is correct.

> **Faster, more reliable testing**
> Traditionally, testing user stories involved slower integration tests in a test environment. The hexagonal architecture enables unit tests to replace some of these integration tests, speeding up our builds and providing greater repeatability of our testing.

We can now test-drive at three granularities against our domain model:

- Against a single method or function

- Against the public behaviors of a class and any collaborators it has

- Against the core logic of an entire user story

This is a big benefit of the hexagonal architecture. The isolation from external services has the effect of pushing the essential logic of a user story into the domain model, where it interacts with ports. As we've seen, those ports – by design – are trivially easy to write test doubles for. It's worth restating the key benefits of FIRST unit tests:

- They are very fast, so testing our user stories will be very fast

- They are highly repeatable, so we can trust test passes and failures

As we cover wide areas of functionality with unit tests, we blur the line between integration and unit testing. We remove friction from developers testing more of the user stories by making that testing easier. Using more unit tests improves build times, as the tests run quickly and give reliable pass/ fail results. Fewer integration tests are needed, which is good as they run more slowly and are more prone to incorrect results.

In the next section, we'll apply what we've learned to our Wordz application. We will write a port that abstracts out the details of fetching a word for our users to guess.

Wordz – abstracting the database

In this section, we will apply what we've learned to our Wordz application and create a port suitable for fetching the words to present to a user. We will write the adapters and integration tests in *Chapter 14, Driving the Database Layer*.

Designing the repository interface

The first job in designing our port is to decide what it should be doing. For a database port, we need to think about the split between what we want our domain model to be responsible for and what we will push out to the database. The ports we use for a database are generally called repository interfaces.

Three broad principles should guide us:

- Think about what the domain model needs – why do we need this data? What will it be used for?

- Don't simply echo an assumed database implementation – don't think in terms of tables and foreign keys at this stage. That comes later when we decide how to implement the storage. Sometimes, database performance considerations mean we have to revisit the abstraction we create here. We would then trade off leaking some database implementation details here if it allowed the database to function better. We should defer such decisions as late as we can.

- Consider when we should leverage the database engine more. Perhaps we intend to use complex stored procedures in the database engine. Reflect this split of behavior in the repository interface. It may suggest a higher-level abstraction in the repository interface.

For our running example application, let's consider the task of fetching a word at random for the user to guess. How should we divide the work between the domain and database? There are two broad options:

- Let the database choose a word at random

- Let the domain model generate a random number and let the database supply a numbered word

In general, letting the database do more work results in faster data handling; the database code is closer to the data and isn't dragging it over a network connection into our domain model. But how do we persuade a database to choose something at random? We know that for relational databases, we can issue a query that will return results in no guaranteed order. That's sort of random. But would it be random enough? Across all possible implementations? Seems unlikely.

The alternative is to let the domain model code decide which word to pick by generating a random number. We can then issue a query to fetch the word associated with that number. This also suggests that each word has an associated number with it – something we can provide when we design the database schema later.

This approach implies we need the domain model to pick a random number from all the numbers associated with the words. That implies the domain model needs to know the full set of numbers to choose from. We can make another design decision here. The numbers used to identify a word will start at 1 and increase by one for each word. We can provide a method on our port that returns the upper bound of these numbers. Then, we are all set to define that repository interface – with a test.

The test class starts with the package declaration and library imports we need:

```
package com.wordz.domain;


import org.junit.jupiter.api.BeforeEach;
import org.junit.jupiter.api.Test;
import org.junit.jupiter.api.extension.ExtendWith;
```

```
import org.mockito.Mock;
import org.mockito.MockitoAnnotations;

import static org.assertj.core.api.Assertions.*;
import static org.mockito.Mockito.when;
```

We enable Mockito integration with an annotation provided by the `junit-jupiter` library. We add the annotation at the class level:

```
@ExtendWith(MockitoExtension.class)
public class WordSelectionTest {
```

This will ensure that Mockito is initialized on each test run. The next part of the test defines some integer constants for readability:

```
private static final int HIGHEST_WORD_NUMBER = 3;
private static final int WORD_NUMBER_SHINE = 2;
```

We need two test doubles, which we want Mockito to generate. We need a stub for the word repository and a stub for a random number generator. We must add fields for these stubs. We will mark the fields with the Mockito `@Mock` annotation so that Mockito will generate the doubles for us:

```
@Mock
private WordRepository repository;

@Mock
private RandomNumbers random;
```

Mockito sees no difference between a mock or stub when we use the `@Mock` annotation. It simply creates a test double that can be configured for use either as a mock or a stub. This is done later in the test code.

We will name the test method `selectsWordAtRandom()`. We want to drive out a class that we will call `WordSelection` and make it responsible for choosing one word at random from `WordRepository`:

```
@Test
void selectsWordAtRandom() {
    when(repository.highestWordNumber())
        .thenReturn(HIGHEST_WORD_NUMBER);
```

```
when(repository.fetchWordByNumber(WORD_NUMBER_SHINE))
    .thenReturn("SHINE");

when(random.next(HIGHEST_WORD_NUMBER))
    .thenReturn(WORD_NUMBER_SHINE);

var selector = new WordSelection(repository,
                                 random);

String actual = selector.chooseRandomWord();

assertThat(actual).isEqualTo("SHINE");
    }
}
```

The preceding test was written in the normal way, adding lines to capture each design decision:

- The WordSelection class encapsulates the algorithm, which selects a word to guess

- The WordSelection constructor takes two dependencies:

  - WordRepository is the port for stored words

  - RandomNumbers is the port for random number generation

- The chooseRandomWord() method will return a randomly chosen word as a String

- The arrange section is moved out to the beforeEachTest() method:

  ```
  @BeforeEach
  void beforeEachTest() {
      when(repository.highestWordNumber())
                  .thenReturn(HIGHEST_WORD_NUMBER);

      when(repository.fetchWordByNumber(WORD_NUMBER_SHINE))
                  .thenReturn("SHINE");
  }
  ```

This will set up the test data in the stub for our WordRepository at the start of each test. The word identified by number 2 is defined as SHINE, so we can check that in the assert.

- Out of that test code flows the following definition of two interface methods:

```
package com.wordz.domain;

public interface WordRepository {
    String fetchWordByNumber(int number);
    int highestWordNumber();
}
```

The WordRepository interface defines our application's view of the database. We only need two facilities for our current needs:

- A fetchWordByNumber() method to fetch a word, given its identifying number
- A highestWordNumber() method to say what the highest word number will be

The test has also driven out the interface needed for our random number generator:

```
package com.wordz.domain;

public interface RandomNumbers {
    int next(int upperBoundInclusive);
}
```

The single next() method returns int in the range of 1 to the upperBoundInclusive number.

With both the test and port interfaces defined, we can write the domain model code:

```
package com.wordz.domain;

public class WordSelection {
    private final WordRepository repository;
    private final RandomNumbers random;

    public WordSelection(WordRepository repository,
                         RandomNumbers random) {
        this.repository = repository;
        this.random = random;
    }

    public String chooseRandomWord() {
```

```
        int wordNumber =
            random.next(repository.highestWordNumber());

        return repository.fetchWordByNumber(wordNumber);
    }
}
```

Notice how this code does not import anything from outside the `com.wordz.domain` package. It is pure application logic, relying only on the port interfaces to access stored words and random numbers. With this, our production code for the domain model of `WordSelection` is complete.

Designing the database and random numbers adapters

The next job is to implement the `RandomNumbers` port and database access code that implements our `WordRepository` interface. In outline, we'll choose a database product, research how to connect to it and run database queries, then test-drive that code using an integration test. We will defer doing these tasks to part three of this book, in *Chapter 13*, *Driving the Domain Layer*, and *Chapter 14*, *Driving the Database Layer*.

Summary

In this chapter, we learned how to apply the SOLID principles to decouple external systems completely, leading to an application architecture known as the hexagonal architecture. We saw how this allows us to use test doubles in place of external systems, making our tests simpler to write, with repeatable results. This, in turn, allows us to test entire user stories with a FIRST unit test. As a bonus, we isolate ourselves from future changes in those external systems, limiting the amount of rework that would be required to support new technologies. We've seen how the hexagonal architecture combined with dependency injection allows us to support several different external systems choices and select the one we want at runtime via configuration.

The next chapter will look at the different styles of automated testing that apply to the different sections of a hexagonal architecture application. This approach is summarized as the Test Pyramid, and we shall learn more about it there.

Questions and answers

Take a look at the following questions and answers regarding this chapter's content:

1. Can we add the hexagonal architecture later?

    Not always. We can refactor it. The challenge can be too much code that directly depends on details of external systems. If that's the starting point, this refactoring will be challenging. There will be a lot of rework to do. This implies that some degree of up-front design and architectural discussion is required before we start work.

2. Is the hexagonal architecture specific to OOP?

    No. It is a way of organizing dependencies in our code. It can be applied to OOP, FP, procedural programming, or anything else – so long as those dependencies are managed correctly.

3. When should we not use the hexagonal architecture?

    When we have no real logic in our domain model. This is common for very small CRUD microservices that typically frontend a database table. With no logic to isolate, putting in all this code has no benefit. We may as well do TDD with integration tests only and accept that we won't be able to use FIRST unit tests.

4. Can we only have one port for an external system?

    No. It is often better if we have more ports. Suppose we have a single Postgres database connected to our application, holding data on users, sales, and product inventory. We could simply have a single repository interface, with methods to work with those three datasets. But it will be better to split that interface up (following ISP) and have `UserRepository`, `SalesRepository`, and `InventoryRepository`. The ports provide a view of what our domain model wants from external systems. Ports are not a one-to-one mapping to hardware.

Further reading

To learn more about the topics that were covered in this chapter, take a look at the following resources:

-   Hexagonal architecture, Alastair Cockburn: `https://alistair.cockburn.us/hexagonal-architecture/`

    The original description of the hexagonal architecture in terms of ports and adapters.

-   `https://medium.com/pragmatic-programmers/unit-tests-are-first-fast-isolated-repeatable-self-verifying-and-timely-a83e8070698e`

    Credits the original inventors of the term FIRST, Tim Ottinger and Brett Schuchert.

-   `https://launchdarkly.com/blog/testing-in-production-for-safety-and-sanity/`

    Guide to testing code deployed on production systems, without accidentally triggering unintended consequences.

FIRST Tests and the Test Pyramid

So far in this book, we've seen the value of writing unit tests that run quickly and give repeatable results. Called FIRST tests, these provide rapid feedback on our design. They are the gold standard of unit tests. We've also seen how the hexagonal architecture helps us design our code in a way that gets the maximum amount covered by FIRST unit tests. But we've also limited ourselves to testing only our domain model – the core of our application logic. We simply have no tests covering how our domain model behaves once it connects to the outside world.

In this chapter, we will cover all the other kinds of tests that we need. We will introduce the test pyramid, which is a way of thinking about the different kinds of tests needed, and how many of each we should have. We'll discuss what each kind of test covers and useful techniques and tools to help. We'll also bring the whole system together by introducing CI/CD pipelines and test environments, outlining the critical role they play in combining code components to create a system for our end users.

In this chapter, we're going to cover the following main topics:

- The test pyramid
- Unit tests – FIRST tests
- Integration tests
- End-to-end and user acceptance tests
- CI/CD pipelines and test environments
- Wordz – integration test for our database

Technical requirements

The code for this chapter can be found at https://github.com/PacktPublishing/Test-Driven-Development-with-Java/tree/main/chapter10.

To run this code, we will need to install the open source Postgres database locally.

To install Postgres, do the following:

1. Go to `https://www.postgresql.org/download/` in your browser.

2. Click on the correct installer for your operating system:

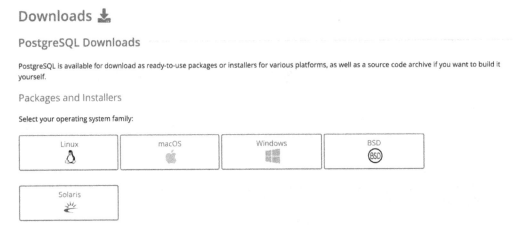

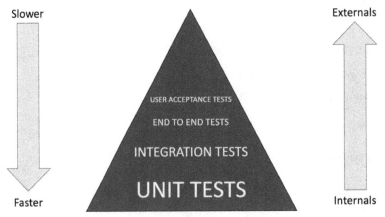

Figure 10.1 – Postgres installer selection

3. Follow the instructions for your operating system.

The test pyramid

A very useful way of thinking about different kinds of tests is by using the **test pyramid**. It is a simple graphical representation of the different kinds of tests we need around our code and the relative numbers of each. This section introduces the key ideas behind the test pyramid.

The test pyramid in graphic form looks as follows:

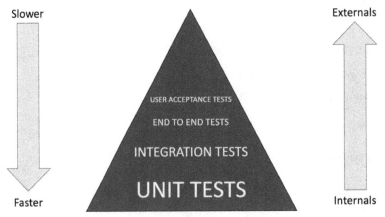

Figure 10.2 – The test pyramid

We can see from the previous graphic that tests are divided into four layers. We have unit tests at the bottom. Integration tests are layered on top of those. The pyramid is completed by end-to-end and user acceptance tests at the top. The graphic shows unit tests in our system are the highest in number, with fewer integration tests and the least number of acceptance tests.

Some of these kinds of tests are new to this book. Let's define what they are:

- **Unit tests**

  These are familiar. They are the FIRST tests we have been using up until now. One defining feature of these tests is that they do not require the presence of any external systems, such as databases or payment processors.

- **Integration tests**

  These tests verify that a software component is correctly integrated with an external system, such as a database. These tests are slow and critically dependent on the external environment being available and correctly set up for our test.

- **End-to-end tests**

  These are the broadest of all tests. An end-to-end test represents something very close to the end user experience. This test is performed against all the real components of the system – possibly in test environments with test data – using the same commands as a real user would use.

- **User acceptance tests**

  This is where the real system is tested as a user would use it. Here, we can confirm that the final system is fit for purpose, according to the requirements the user has given us.

It's not obvious at first why having fewer tests of any kind would be an advantage. After all, everything up until now in this book has positively praised the value of testing. Why do we not simply have *all the tests*? The answer is a pragmatic one: not all tests are created equal. They don't all offer equal value to us as developers.

The reason for the shape of this pyramid is to reflect the practical value of each layer of testing. Unit tests written as FIRST tests are *fast* and *repeatable*. If we could build a system out of only these unit tests, we surely would. But unit tests do not exercise every part of our code base. Specifically, they do not exercise connections from our code to the outside world. Nor do they exercise our application in the same way as a user would use it. As we progress up through the layers of testing, we move away from testing the *internal* components of our software and move toward testing how it interacts with *external* systems and, ultimately, the end user of our application.

The test pyramid is about *balance*. It aims to create layers of tests that achieve the following:

- Run as quickly as possible
- Cover as much code as possible

- Prevent as many defects as possible

- Minimize duplication of the test effort

In the following sections, we will look at a breakdown of the tests involved at each layer of the test pyramid. We'll consider the strengths and weaknesses of each kind of test, allowing us to understand what the test pyramid is guiding us toward.

Unit tests – FIRST tests

In this section, we're going to look at the base of the test pyramid, which consists of unit tests. We'll examine why this layer is critical to success.

By now, we're very familiar with FIRST unit tests. The preceding chapters have covered these in detail. They are the gold standard of unit tests. They are fast to run. They are repeatable and reliable. They run isolated from each other, so we can run one, or run many and run them in any order we choose. FIRST tests are the powerhouses of TDD, enabling us to work with a rapid feedback loop as we code. Ideally, all our code would fall under this feedback loop. It provides a fast, efficient way to work. At every step, we can execute code and prove to ourselves that it is working as we intended. As a helpful byproduct, by writing tests that exercise each possible desirable behavior in our code, we will end up exercising every possible code path. We will get 100% *meaningful* test coverage of code under unit tests when we work in this way.

Because of their advantages, unit tests form the bedrock of our testing strategy. They are represented as the base of the test pyramid.

Unit tests have advantages and limitations, as summarized in the following table:

Advantages	Limitations
These are the fastest-running tests and provide the fastest possible feedback loop for our code.	The smaller scope of these tests means that having all unit tests pass is no guarantee that the system as a whole is working correctly.
Stable and repeatable, having no dependencies on things outside of our control.	They can be written with too strong a tie to implementation details, making future additions and refactoring difficult.
Can provide very detailed coverage of a specific set of logic. Locate defects accurately.	Not helpful for testing interactions with external systems.

Table 10.1 – Unit test advantages and disadvantages

In any system, we expect to have the largest number of tests at the unit level. The test pyramid represents this graphically.

We can't achieve full coverage by using unit tests alone in the real world but we can improve our situation. By applying the hexagonal architecture to our application, we can get the majority of code under unit tests. Our fast-running unit tests can cover a lot of ground like this and provide a lot of confidence in our application logic. We can get as far as knowing that if the external systems behave as we expect them to, our domain layer code will be able to correctly handle every use case we have thought about.

The test position when using unit tests alone is shown in the following diagram:

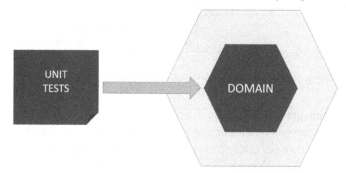

Figure 10.3 – Unit tests cover the domain model

Unit tests only test components of our domain model. They do not test external systems, nor do they use external systems. They rely on test doubles to simulate our external systems for us. This gives us advantages in development cycle speed but has the drawback that our connections to those external systems remain untested. If we have a piece of unit-tested code that accesses a repository interface, we know that its logic works with a stub repository. Its internal logic will even have 100% test coverage and this will be valid. But we won't know if it will work with the real repository yet.

The adapter layer code is responsible for those connections, and it is not tested at the unit test level. To test this layer, we're going to need a different approach to testing. We will need to test what happens when our domain layer code is integrated with actual external systems.

The next section looks at how we test these external systems adapters using a kind of testing known as integration tests.

Integration tests

In this section, we're going to look at the next layer up in the test pyramid: integration testing. We'll see why it's important, review helpful tools, and understand the role of integration testing in the overall scheme of things.

Integration tests exist to test that our code will successfully integrate with external systems. Our core application logic is tested by unit tests, which, by design, do not interact with external systems. This means that we need to test behavior with those external systems at some point.

Integration tests are the second layer up in the test pyramid. They have advantages and limitations, as summarized in the following table:

Advantages	Limitations
Test that software components interact correctly when connected	Require test environments to be set up and maintained
Provide a closer simulation of the software system as it will be used live	Tests run more slowly than unit tests
	Susceptible to problems in the test environment, such as incorrect data or network connection failures

Table 10.2 – Integration test advantages and disadvantages

There should be fewer integration tests than unit tests. Ideally, far fewer. While unit tests avoided many problems of testing external systems by using test doubles, integration tests must now face those challenges. By nature, they are more difficult to set up. They *can* be less repeatable. They generally run more slowly than unit tests do, as they wait for responses from external systems.

To give a sense of this, a typical system might have thousands of unit tests and hundreds of acceptance tests. In between, we have several integration tests. Many integration tests point to a design opportunity. We can refactor the code so that our integration test is pushed down to being a unit test or promoted to being an acceptance test.

Another reason to have fewer integration tests is due to **flaky tests**. A flaky test is a nickname given to a test that sometimes passes and sometimes fails. When it fails, it is due to some problem interacting with the external system and not a defect in the code we are testing. Such a failure is called a **false negative** test result – a result that can mislead us.

Flaky tests are a nuisance precisely because we cannot immediately tell the root cause of the failure. Without diving into error logs, we only know that the test failed. This leads to developers learning to ignore these failed tests, often choosing to re-run the test suite several times until the flaky test passes. The problem here is that we are training developers to have less faith in their tests. We are training them to ignore test failures. This is not a good place to be.

What should an integration test cover?

In our design so far, we have decoupled external systems from our domain code using the *Dependency Inversion Principle*. We have created an interface defining how we use that external system. There will be some implementation of this interface, which is what our integration test will be covering. In hexagonal architecture terms, this is an *adapter*.

This adapter should only contain the minimum amount of code necessary to interact with the external system in a way that satisfies our interface. It should have no application logic in it at all. That should be inside the domain layer and covered by unit tests. We call this a *thin adapter*, doing only enough work to adapt to the external system. This means our integration test is nicely limited in scope.

We can represent the scope of an integration test like so:

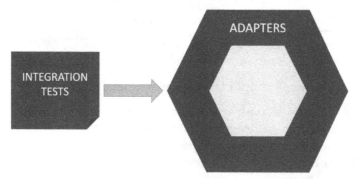

Figure 10.4 – Integration tests cover the adapter layer

Integration tests only test the adapter layer components, those pieces of code that directly interact with external systems, such as databases and web endpoints. The integration test will create an instance of the adapter under test and arrange for it to connect to *a version* of the external service. This is important. We're still not connecting to the production services yet. Until the integration test passes, we're not sure that our adapter code works correctly. So, we don't want to access real services just yet. We also want to have that extra level of control over these services. We want to be able to safely and easily create test accounts and fake data to use with our adapter. That means we need a collection of live-like services and databases to use. That means they have to live and run somewhere.

Test environments are the name given to the arrangement of external systems we use in integration tests. It is an environment for running web services and data sources, specifically for testing.

A test environment enables our code to connect to test versions of real external systems. It's one step closer to production readiness, compared to the unit test level. There are some challenges involved in using test environments, however. Let's look into the good practices for testing integrations with databases and web services.

Testing database adapters

The basic approach to testing a database adapter is to set up a database server in the test environment and get the code under test to connect to it. The integration test will preload a known dataset into the database as part of its Arrange step. The test then runs the code that interacts with the database in the Act step. The Assert step can inspect the database to see if expected database changes happened.

The biggest challenge in testing a database is that it remembers data. Now, this might seem a little obvious, as that is the entire point of using a database in the first place. But it conflicts with one of the goals of testing: to have isolated, repeatable tests. As an example, if our test created a new user account for user `testuser1` and that was stored in the database, we would have a problem running that test again. It would not be able to create `testuser1` and instead would receive a **user already exists** error.

There are different approaches to overcoming this problem, each with trade-offs:

- **Delete all data from the database before and after each test case**

  This approach preserves the isolation of our tests, but it is slow. We have to recreate the test database schema before every test.

- **Delete all data before and after the full set of adapter tests run**

  We delete data less often, allowing several related tests to run against the same database. This loses test isolation due to the stored data, as the database will not be in the state expected at the start of the next test. We have to run tests in a particular order, and they must all pass, to avoid spoiling the database state for the next test. This is not a good approach.

- **Use randomized data**

  Instead of creating *testuser1* in our test, we randomize names. So, on one run, we might get `testuser-cfee-0a9b-931f`. On the next run, the randomly chosen username would be something else. The state stored in the database will not conflict with another run of the same test. This is another way to preserve test isolation. However, it does mean that tests can be harder to read. It requires periodic cleanup of the test database.

- **Rollback transactions**

  We can add data required by our tests inside a database transaction. We can roll back the transaction at the end of the test.

- **Ignore the problem**

  Sometimes, if we work with read-only databases, we can add test data that will never be accessed by the production code and leave it there. If this works, it is an attractive option requiring no extra effort.

Tools such as *database-rider*, available from `https://database-rider.github.io/getting-started/`, assist by providing library code to connect to databases and initialize them with test data.

Testing web services

A similar approach is used to test the integration with web services. A test version of the web service is set to run in the test environment. The adapter code is set to connect to this test version of the web service, instead of the real version. Our integration test can then examine how the adapter code behaves. There might be additional web APIs on the test service to allow inspection by the assertions in our test.

Again, the disadvantages are a slower running test and the risk of flaky tests due to issues as trivial as network congestion.

Sandbox APIs

Sometimes, hosting our own local service might be impossible, or at least undesirable. Third-party vendors are usually unwilling to release test versions of their service for us to use in our test environment. Instead, they typically offer a **sandbox API**. This is a version of their service that the third party hosts, not us. It is disconnected from their production systems. This sandbox allows us to create test accounts and test data, safe from affecting anything real in production. It will respond to our requests as their production versions will respond, but without taking any action such as taking payment. Consider them test simulators for real services.

Consumer-driven contract testing

A useful approach to testing interactions with web services is called **consumer-driven contract testing**. We consider our code as having a contract with the external service. We agree to call certain API functions on the external service, supplying data in the form required. We need the external service to respond to us predictably, with data in a known format and well-understood status codes. This forms a *contract* between the two parties – our code and the external service API.

Consumer-driven contract testing involves two components, based on that contract, often using code generated by tools. This is represented in the following figure:

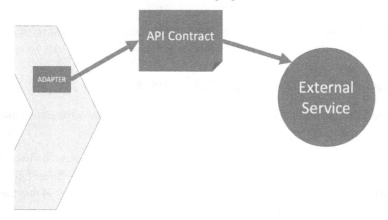

Figure 10.5 – Consumer-driven contract testing

The preceding diagram shows that we've captured the expected interactions with an external service as an API contract. Our adapter for that service will be coded to implement that API contract. When using consumer-driven contract testing, we end up with two tests, which test either side of that contract. If we consider a service to be a black box, we have a public interface presented by the black box, and an implementation, whose details are hidden inside that black box. A contract test is two tests. One test confirms that the outside interface is compatible with our code. The other test confirms that the implementation of that interface works and gives the expected results.

A typical contract test will need two pieces of code:

- **A stub of the external service**: A stub of the external service is generated. If we are calling a payment processor, this stub simulates the payment processor locally. This allows us to use it as a test double for the payment processor service as we write our adapter code. We can write an integration test against our adapter, configuring it to call this stub. This allows us to test our adapter code logic without accessing the external system. We can verify that the adapter sends the correct API calls to that external service and handles the expected responses correctly.

- **A replay of a set of calls to the real external service**: The contract also allows us to run tests against the real external service – possibly in sandbox mode. We're not testing the functionality of the external service here – we assume that the service provider has done that. Instead, we are verifying that what we believe about its API is true. Our adapter has been coded to make certain API calls in certain orders. This test verifies that this assumption is correct. If the test passes, we know that our understanding of the external service API was correct and also that it has not changed. If this test was previously working but now fails, that would be an early indication that the external service has changed its API. We would then need to update our adapter code to follow that.

One recommended tool for doing this is called Pact, available at `https://docs.pact.io`. Read the guides there for more details on this interesting technique.

We've seen that integration tests get us one step nearer to production. In the next section, we look at the final level of testing in the test pyramid, which is the most live-like so far: user acceptance tests.

End-to-end and user acceptance tests

In this section, we will progress to the top of the test pyramid. We'll review what end-to-end and user acceptance tests are and what they add to unit and integration testing.

At the top of the test pyramid lies two similar kinds of tests called **end-to-end tests** and **user acceptance tests**. Technically, they are the same kind of test. In each case, we start up the software fully configured to run in its most live-like test environment, or possibly in production. The idea is that the system is tested as a whole from one end to the other.

One specific use of an end-to-end test is for **user acceptance testing** (UAT). Here, several key end-to-end test scenarios are run. If they all pass, the software is declared fit for purpose and accepted by the users. This is often a contractual stage in commercial development, where the buyer of the software formally agrees that the development contract has been satisfied. It's still end-to-end testing that is being used to determine that, with cherry-picked test cases.

These tests have advantages and limitations, as summarized in the following table:

Advantages	Limitations
Most comprehensive testing of functionality available. We are testing at the same level that a user of our system – either person or machine – would experience our system.	Slowest tests to run.
Tests at this level are concerned with pure behavior as observed from outside the system. We could refactor and rearchitect large parts of the system and still have these tests protect us.	Reliability issues – many problems in the setup and environment of our system can cause false negative test failures. This is termed "brittleness" – our tests are highly dependent on their environment working correctly. Environments can be broken due to circumstances beyond our control.
Contractually important – these tests are the essence of what the end user cares about.	These are the most challenging of all the tests to write, due to the extensive environment setup requirements.

Table 10.3 – End-to-end test advantages and disadvantages

Acceptance tests having a spot at the top of the pyramid is a reflection that we don't need many of them. The majority of our code should now be covered by unit and integration tests, assuring us that our application logic works, as well as our connections to external systems.

The obvious question is *what's left to test*? We don't want to duplicate testing that has already been done at the unit and integration levels. But we do need some way to validate that the software *as a whole* is going to work as expected. This is the job of end-to-end testing. This is where we configure our software so that it connects to real databases and real external services. Our production code has passed all the unit tests with test doubles. These test passes suggest our code *should* work when we connect these real external services. But *should* is a wonderful weasel word in software development. Now, is the time to verify that it does, using an end-to-end test. We can represent the coverage of these tests using the following diagram:

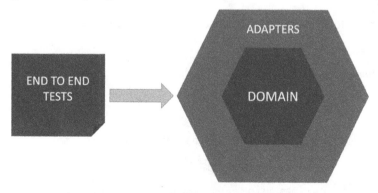

Figure 10.6 – End-to-end/user acceptance tests cover the entire code base

End-to-end tests cover the entire code base, both the domain model and the adapter layer. As such, it repeats testing work already done by unit and integration tests. The main technical aspect we want to test in end-to-end testing is that our software is configured and wired up correctly. Throughout this book, we have used *dependency inversion and injection* to isolate us from external systems. We've created test doubles and injected those. Now, we must create actual production code, the real adapter layer components that connect to the production systems. We inject those into our system during its initialization and configuration. This sets the code up to work for real.

End-to-end tests will then duplicate a *small amount* of happy path testing already covered by unit and integration tests. The purpose here is *not* to verify the behaviors that we have already tested. Instead, these tests verify that we have injected the correct production objects, by confirming that the system as a whole behaves correctly when connected to production services.

A user acceptance test builds on this idea by running through key test scenarios considered critical to accepting the software as complete. These will be end-to-end tests at a technical level. But their purpose is broader than the technical goal of ensuring our system is correctly configured. They are more of a legal contractual nature: *Have we built what was asked of us?* By using the iterative approach in this book together with its technical practices, there's a higher chance that we will have done so.

Acceptance testing tools

Various testing libraries exist to help us write automated acceptance and end-to-end tests. Tasks such as connecting to a database or calling an HTTP web API are common to this kind of testing. We can leverage libraries for these tasks, instead of writing code ourselves.

The main differentiator among these tools is the way they interact with our software. Some are intended to simulate a user clicking a desktop GUI, or a browser-based web UI. Others will make HTTP calls to our software, exercising a web endpoint.

Here are a few popular acceptance testing tools to consider:

- **RestEasy**

  A popular tool for testing REST APIs: `https://resteasy.dev/`

- **RestAssured**

  Another popular tool for testing REST APIs that takes a fluent approach to inspecting JSON responses: `https://rest-assured.io/`

- **Selenium**

  A popular tool for testing web UIs through the browser: `https://www.selenium.dev/`

- **Cucumber**

  Available from `https://cucumber.io/`. Cucumber allows English language-like descriptions of tests to be written by domain experts. At least, that's the theory. I've never seen anybody other than a developer write Cucumber tests in any project I've been part of.

Acceptance tests form the final piece of the test pyramid and allow our application to be tested under conditions that resemble the production environment. All that is needed is a way to automate running all those layers of testing. That's where CI/CD pipelines come in, and they are the subject of the next section.

CI/CD pipelines and test environments

CI/CD pipelines and test environments are an important part of software engineering. They are a part of the development workflow that takes us from writing code to having systems in the hands of users. In this section, we're going to look at what the terms mean and how we can use these ideas in our projects.

What is a CI/CD pipeline?

Let's start with defining the terms:

- CI stands for **continuous integration**

  Integration is where we take individual software components and join them together to make a whole. CI means we do this all the time as we write new code.

- CD stands for either **continuous delivery** or **continuous deployment**

  We'll cover the difference later, but in both cases, the idea is that we are taking the latest and greatest version of our integrated software and delivering it to a stakeholder. The goal of continuous delivery is that we could – if we wanted to – deploy every single code change to production with a single click of a button.

It's important to note that CI/CD is an engineering *discipline* – not a set of tools. However we achieve it, CI/CD has the goal of growing a single system that is always in a usable state.

Why do we need continuous integration?

In terms of the test pyramid, the reason we need CI/CD is to pull all the testing together. We need a mechanism to build the whole of our software, using the latest code. We need to run all the tests and ensure they all pass before we can package and deploy the code. If any tests fail, we know the code is not suitable for deployment. To ensure we get fast feedback, we must run the tests in order of fastest to slowest. Our CI pipeline will run unit tests first, followed by integration tests, followed by end-to-end and acceptance tests. If any tests fail, the build will produce a report of test failures for that stage, then stop the build. If all the tests pass, we package our code up ready for deployment.

More generally, the idea of **integration** is fundamental to building software, whether we work alone or in a development team. When working alone, following the practices in this book, we're building software out of several building blocks. Some we have made ourselves, while for others, we've selected a suitable library component and used that. We've also written adapters – components that allow us to access external systems. All of that needs integrating – bringing together as a whole – to turn our lines of code into a working system.

When working in a team, integration is even more important. We need to not only bring together the pieces we have written but also all the other pieces written by the rest of our team. Integrating work in progress from colleagues is urgent. We end up building on what others have already written. As we work outside of the main integrated code base, there is a risk of not including the latest design decisions and pieces of reusable code.

The following figure shows the goal of CI:

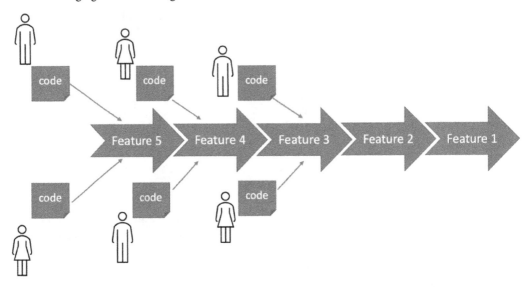

Figure 10.7 – Continuous integration

The motivation behind CI was to avoid the classic waterfall development trap, where a team wrote code as isolated individuals while following a plan and only integrated it at the end. Many times, that integration failed to produce working software. There was often some misunderstanding or missing piece that meant components did not fit together. At this late stage of a waterfall project, mistakes are expensive to fix.

It's not just big teams and big projects that suffer from this. My turning point was while writing a flight simulator game for Britain's RAF Red Arrows display team. Two of us worked on that game to a common API we had agreed on. When we first attempted to integrate our parts – at 03:00 A.M., in front of the company managing director, of course – the game ran for about three frames and then crashed. Oops! Our lack of CI provided an embarrassing lesson. It would have been good to know that was going to happen a lot earlier, especially without the managing director watching.

Why do we need continuous delivery?

If CI is all about keeping our software components together as an ever-growing whole, then CD is about getting that whole into the hands of people who care about it. The following figure illustrates CD:

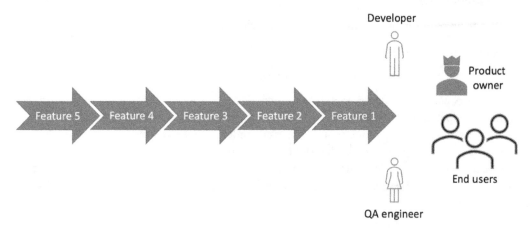

Figure 10.8 – Continuous delivery

Delivering a stream of value to end users is a core tenet of agile development. No matter which flavor of agile methodology you use, getting features into the hands of users has always been the goal. We want to deliver usable features at regular, short intervals. Doing this provides three benefits:

- **Users get the value they want**

  End users don't care about our development process. They only care about getting solutions to their problems. Whether that's the problem of being entertained while waiting for an Uber ride, or the problem of paying everyone's wages in a multinational business, our user just wants their problem gone. Getting valuable features to our users becomes a competitive advantage.

- **We gain valuable user feedback**

  Yes, that's what I asked for – but it isn't what I meant! That is extremely valuable user feedback that agile approaches deliver. Once an end user sees the feature as we have implemented it, sometimes, it becomes clear that it isn't quite solving their problem. We can correct this quickly.

- **Aligns the code base and development team**

  To pull off this feat, you do need to have your team and workflows together. You can't effectively do this unless your workflow results in known working software being continuously available as a single whole.

Continuous delivery or continuous deployment?

Exact definitions of these terms seem to vary, but we can think of them like this:

- **Continuous delivery**

  We deliver software to internal stakeholders, such as product owners and QA engineers

- **Continuous deployment**

  We deliver software into production and to end users

Out of the two, continuous *deployment* sets a much higher bar. It requires that once we integrate code into our pipeline, that code is ready to go live – into production, to real users. This is, of course, hard. It needs top-class test automation to give us confidence that our code is ready to deploy. It also benefits from having a fast rollback system in production – some means of quickly reverting a deployment if we discover a defect not covered by our tests. Continuous deployment is the ultimate workflow. For all who achieve it, deploying new code last thing on Friday simply holds no fear. Well, maybe a little less fear.

Practical CI/CD pipelines

Most projects use a CI tool to handle the sequencing chores. Popular tools are provided by Jenkins, GitLab, CircleCI, Travis CI, and Azure DevOps. They all work similarly, executing separate build stages sequentially. That's where the name pipeline comes from – it resembles a pipe being loaded at one end with the next build stage and coming out of the other end of the pipe, as shown in the following diagram:

Figure 10.9 – Stages in a CI pipeline

A CI pipeline comprises the following steps:

1. **Source control**: Having a common location in which to store the code is essential to CI/CD. It is the place where code gets integrated. The pipeline starts here, by pulling down the latest version of the source code and performing a clean build. This prevents errors caused by older versions of code being present on the computer.

2. **Build**: In this step, we run a build script to download all the required libraries, compile all the code, and link it together. The output is something that can be executed, typically a single Java archive .jar file, to run on the JVM.

3. **Static code analysis**: Linters and other analysis tools check the source code for stylistic violations, such as variable length and naming conventions. The development team can choose to fail the build when specific code issues are detected by static analysis.

4. **Unit tests**: All the unit tests are run against the built code. If any fail, the pipeline stops. Test failure messages are reported.

5. **Integration tests**: All integration tests are run against the built code. If any fail, the pipeline is stopped and error messages are reported.

6. **Acceptance tests**: All acceptance tests are run against the built code. If all tests pass, the code is considered to be working and ready for delivery/deployment.

7. **Delivery packaging**: The code is packaged up into a suitable form for delivery. For Java web services, this may well be a single Java archive `.jar` file containing an embedded web server.

What happens next depends on the needs of the project. The packaged code may be deployed to production automatically or it may simply be placed in some internal repository, for access by product owners and QA engineers. Formal deployment would then happen later, after quality gatekeeping.

Test environments

One obvious problem caused by needing a CI pipeline to run integration tests is having a place to run those tests. Ordinarily, in production, our application integrates with external systems such as databases and payment providers. When we run our CI pipeline, we do not want our code to process payments or write to production databases. Yet we *do* want to test that the code *could* integrate with those things, once we configure it to connect to those real systems.

The solution is to create a **test environment**. These are collections of databases and simulated external systems that lie under our control. If our code needs to integrate with a database of user details, we can create a copy of that user database and run it locally. During testing, we can arrange for our code to connect to this local database, instead of the production version. External payment providers often provide a sandbox API. This is a version of their service that, again, does not connect to any of their real customers. It features simulated behavior for their service. In effect, it is an external test double.

This kind of setup is called a **live-like** or **staging** environment. It allows our code to be tested with more realistic integration. Our unit tests use stubs and mocks. Our integration tests can now use these richer test environments.

Advantages and challenges of using test environments

Test environments offer both advantages and disadvantages, as summarized in the following table:

Advantages	Challenges
The environment is self-contained We can create it and destroy it at will. It will not affect production systems.	**Not production environments** No matter how live-like we make them, these environments are simulations. The risk is that our fake environments give us false positives – tests that pass only because they are using fake data. This can give us false confidence, leading us to deploy code that will fail in production. The real test happens when we set our code live. Always.
More realistic than stubs The environment gets us one step closer to testing under production loads and conditions.	**Extra effort to create and maintain** More development work is needed to set these environments up and keep them in step with the test code.
Check assumptions about external systems Third-party sandbox environments allow us to confirm that our code uses the latest, correct API, as published by the supplier.	**Privacy concerns** Simply copying over a chunk of production data isn't good enough for a test environment. If that data contains **personally identifiable information (PII)** as defined by GDPR or HIPAA, then we can't legally use it directly. We have to create an extra step to anonymize that data or generate pseudo-realistic random test data. Neither is trivial.

Table 10.4 – Test environments advantages and challenges

Testing in production

I can hear the gasps already! Running our tests in production is generally a terrible idea. Our tests might introduce fake orders that our production system treats as real ones. We may have to add test user accounts, which can present a security risk. Worse still, because we are in a testing phase, there is a very good chance that our code does not work yet. This can cause all sorts of problems – all while connected to production systems.

Despite these concerns, sometimes, things must be tested in production. Big data companies such as Google and Meta both have things that can only be tested live due to the sheer scale of their data. There is no way a meaningful live-like test environment can be created; it will simply be too small. What can we do in cases like this?

The approach is to mitigate the risks. Two techniques are valuable here: blue-green deployment and traffic partitioning.

Blue-green deployment

Blue-green deployment is a deployment technique designed for the rapid rollback of failed deployments. It works by dividing the production servers into two groups. They are referred to as *blue* and *green*, chosen as they are neutral colors that both denote success. Our production code will be running on one group of servers at any one time. Let's say we are currently running on the blue group. Our next deployment will then be to the green group. This is shown in the following diagram:

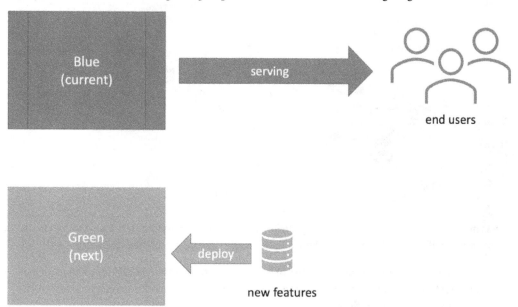

Figure 10.10 – Blue-green deployment

Once the code has been deployed to the green group, we switch over the production configuration to connect to green group servers. We retain the previous working production code on the blue servers. If our testing goes well against the green group, then we're done. Production is now working with the latest green group code. If the testing fails, we revert that configuration to connect to the blue servers once again. It's a fast rollback system that enables our experimentation.

Traffic partitioning

In addition to blue-green deployment, we can limit the amount of traffic that we send to our test servers. Instead of flipping production to wholly use the new code under test, we can simply send a small percentage of user traffic there. So, 99% of users might be routed to our blue servers, which we know to work. 1% can be routed to our new code under test in the green servers, as shown in the following diagram:

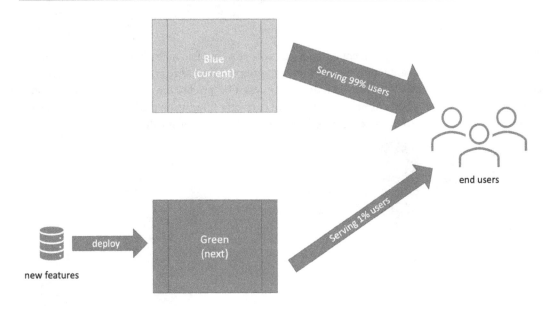

Figure 10.11 – Traffic partitioning

If defects are discovered, only 1% of users will be affected before we revert to 100% blue servers. This gives us a rapid rollback, mitigating problems in production caused by a failed deployment.

We've now covered the roles of different kinds of tests and seen how they fit into a coherent system known as the test pyramid. In the next section, we'll apply some of this knowledge to our Wordz application by writing an integration test.

Wordz – integration test for our database

In this section, we'll review an integration test for our Wordz application to get a feel for what they look like. We'll cover the details of writing these tests and setting up the test tools in *Chapter 14, Driving the Database Layer*, and *Chapter 15, Driving the Web Layer*.

Fetching a word from the database

As part of our earlier design work, we identified that Wordz would need a place to store the candidate words to be guessed. We defined an interface called `WordRepository` to isolate us from the details of storage. At that iteration, we had only got as far as defining one method on the interface:

```
public interface WordRepository {
String fetchWordByNumber( int wordNumber );
}
```

The implementation of this WordRepository interface will access the database and return a word given its wordNumber. We will defer implementing this to *Chapter 14, Driving the Database Layer*. For now, let's take an early look at what the integration test will look like, at a high level. The test uses open source libraries to help write the test, and to provide the database. We've chosen the following:

- An open source library called database-rider (available from https://database-rider.github.io/getting-started/) as a test tool

- Postgres, a popular open source relational database, to store our data

Here is the test code:

```
package com.wordz.adapters.db;

import com.github.database.rider.core.api.connection.
ConnectionHolder;
import com.github.database.rider.core.api.dataset.DataSet;
import com.github.database.rider.junit5.api.DBRider;
import org.junit.jupiter.api.BeforeEach;
import org.junit.jupiter.api.Test;
import org.postgresql.ds.PGSimpleDataSource;

import javax.sql.DataSource;

import static org.assertj.core.api.Assertions.assertThat;

@DBRider
public class WordRepositoryPostgresTest {
    private DataSource dataSource;

    @BeforeEach
    void beforeEachTest() {
        var ds = new PGSimpleDataSource();
        ds.setServerNames(new String[]{"localhost"});
        ds.setDatabaseName("wordzdb");
        ds.setUser("ciuser");
        ds.setPassword("cipassword");

        this.dataSource = ds;
```

```
        }

        private final ConnectionHolder connectionHolder = () ->
            dataSource.getConnection();

        @Test
        @DataSet("adapters/data/wordTable.json")
        public void fetchesWord()  {
            var adapter = new WordRepositoryPostgres(dataSource);

            String actual = adapter.fetchWordByNumber(27);

            assertThat(actual).isEqualTo("ARISE");
        }
    }
```

The fetchesWord() test method is marked by the @DataSet annotation. This annotation is provided by the *database-rider* test framework and it forms the Arrange step of our test. It specifies a file of known test data that the framework will load into the database before the test runs. The data file is located underneath the root folder of src/test/resources. The parameter in the annotation gives the rest of the path. In our case, the file will be located at src/test/resources/adapters/data/wordTable.json. Its content looks like this:

```
{
    "WORD": [
        {
            "id": 1,
            "number": 27,
            "text": "ARISE"
        }
    ]
}
```

This JSON file tells the database-rider framework that we would like to insert a single row into a database table named WORD, with column values of 1, 27, and ARISE.

We're not going to write the adapter code to make this test pass just yet. There are several steps we would need to take to get this test to compile, including downloading various libraries and getting the Postgres database up and running. We'll cover these steps in detail in *Chapter 14, Driving the Database Layer*.

The overview of this integration test code is that it is testing a new class called WordRepositoryPostgres that we will write. That class will contain the database access code. We can see the tell-tale JDBC object, `javax.sql.DataSource`, which represents a database instance. This is the clue that we are testing integration with a database. We can see new annotations from the database testing library: `@DBRider` and `@DataSet`. Finally, we can see something instantly recognizable – the Arrange, Act, and Assert steps of a test:

1. The Arrange step creates a `WordRepositoryPostgres` object, which will contain our database code. It works with the `database-rider` library's `@DataSet` annotation to put some known data into the database before the test runs.

2. The Act step calls the `fetchWordByNumber()` method, passing in the numeric wordNumber we want to test. This number aligns with the contents of the `wordTable.json` file.

3. The Assert step confirms the expected word, `ARISE`, is returned from the database.

As we can see, integration tests aren't so different from unit tests in essence.

Summary

In this chapter, we've seen how the test pyramid is a system that organizes our testing efforts, keeping FIRST unit tests firmly as the foundation for all we do, but not neglecting other testing concerns. First, we introduced the ideas of integration and acceptance testing as ways of testing more of our system. Then, we looked at how the techniques of CI and CD keep our software components brought together and ready to release at frequent intervals. We've seen how to bring the whole build process together using CI pipelines, possibly going on to CD. We've made a little progress on Wordz by writing an integration test for the `WordRepositoryPostgres` adapter, setting us up to write the database code itself.

In the next chapter, we'll take a look at the role of manual testing in our projects. It's clear by now that we automate as much testing as we can, meaning that the role of manual testing no longer means following huge test plans. Yet, manual testing is still very valuable. How has the role changed? We'll review that next.

Questions and answers

The following are some questions and their answers regarding this chapter's material:

1. Why is the test pyramid represented as a pyramid shape?

   The shape depicts a broad foundation of many unit tests. It shows layers of testing above that exercise a closer approximation to the final, integrated system. It also shows that we expect fewer tests at those higher levels of integration.

2. What are the trade-offs between unit, integration, and acceptance tests?

   - Unit tests: Fast, repeatable. Don't test connections to external systems.

- Integration tests: Slower, sometimes unrepeatable. They test the connection to the external system.

- Acceptance tests: Slowest of all. They can be flaky but offer the most comprehensive tests of the whole system.

3. Does the test pyramid guarantee correctness?

No. Testing can only ever reveal the presence of a defect, *never* the absence of one. The value of extensive testing is in how many defects we avoid putting into production.

4. Does the test pyramid only apply to object-oriented programming?

No. This strategy of test coverage applies to any programming paradigm. We can write code using any paradigm - object-oriented, functional, procedural, or declarative. The various kinds of tests only depend on whether our code accesses external systems or makes up purely internal components.

5. Why don't we prefer end-to-end tests, given they test the whole system?

End-to-end tests run slowly. They depend directly on having either production databases and web services running, or a test environment running containing test versions of those things. The network connections required, and things such as database setup, can result in tests giving us false negative results. They fail because of the environment, not because the code was incorrect. Because of these reasons, we engineer our system to make maximum use of fast, repeatable unit tests.

Further reading

To learn more about the topics that were covered in this chapter, take a look at the following resources:

- *Introduction to consumer-driven contract testing*

  Pact.io produce a popular open source contract testing tool that's available on their website, `https://docs.pact.io`. The website features an explanatory video and a useful introduction to the benefits of contract-driven testing.

- *Database-rider database testing library*

  An open source database integration testing library that works with JUnit5. It is available from `https://database-rider.github.io/getting-started/`.

- *Modern Software Engineering, Dave Farley, ISBN 978-0137314911*

  This book explains in detail the reasons behind CD and various technical practices such as trunk-based development to help us achieve that. Highly recommended.

- *Minimum CD*

  Details on what is needed for CD: `https://minimumcd.org/minimumcd/`.

11

Exploring TDD with Quality Assurance

Previous chapters have covered the technical practices needed to design and test well-engineered code. The approach presented has been primarily for developers to gain rapid feedback on software design. Testing has been almost a byproduct of these efforts.

The combination of TDD, continuous integration, and pipelines provides us with a high level of confidence in our code. But they are not the whole picture when it comes to software **Quality Assurance (QA)**. Creating the highest-quality software needs additional processes, featuring the human touch. In this chapter, we will highlight the importance of manual exploratory testing, code reviews, user experience, and security testing, together with approaches to adding a human decision point to a software release.

In this chapter, we're going to cover the following main topics:

- TDD – its place in the bigger quality picture
- Manual exploratory testing – discovering the unexpected
- Code review and ensemble programming
- User interface and user experience testing
- Security testing and operations monitoring
- Incorporating manual elements into CI/CD workflows

TDD – its place in the bigger quality picture

In this section, we will take a critical look at what TDD has brought to the testing table, and what remains human activities. While TDD undoubtedly has advantages as part of a test strategy, it can never be the entire strategy for a successful software system.

Understanding the limits of TDD

TDD is a relatively recent discipline as far as mainstream development goes. The modern genesis of TDD lies with Kent Beck in the Chrysler Comprehensive Compensation System (see the *Further reading* section, where the idea of test-first unit testing came from). The project began in 1993 and Kent Beck's involvement commenced in 1996.

The Chrysler Comprehensive Compensation project was characterized by extensive use of unit tests driving small iterations and frequent releases of code. Hopefully, we recognize those ideas from the preceding chapters in this book. Much has changed since then – the deployment options are different, the number of users has increased, and agile approaches are more common – but the goals of testing remain the same. Those goals are to drive out correct, well-engineered code and ultimately satisfy users.

The alternative to test automation is to run tests without automation – in other words, run them manually. A better term might be human-driven. Before test automation became commonplace, an important part of any development plan was a test strategy document. These lengthy documents defined when testing would be done, how it would be done, and who would be doing that testing.

This strategy document existed alongside detailed test plans. These would also be written documents, describing each test to be performed – how it would be set up, what steps exactly were to be tested, and what the expected results should be. The traditional waterfall-style project would spend a lot of time defining these documents. In some ways, these documents were similar to our TDD test code, only written on paper, rather than source code.

Executing these manual test plans was a large effort. Running a test needs us to set up test data by hand, run the application, then click through user interfaces. Results must be documented. Defects found must be recorded in defect reports. These must be fed back up the waterfall, triggering redesigns and recoding. This must happen with every single release. *Human-driven* testing is repeatable, but only at the great cost of preparing, updating, and following test documents. All of this took time – and a lot of time at that.

Against this background, Beck's TDD ideas seemed remarkable. Test documents became executable code and could be run as often as desired, for a fraction of the cost of human testing. This was a compelling vision. The responsibility of testing code was part of the developer's world now. The tests were part of the source code itself. These tests were automated, capable of being run in full on every build, and kept up to date as the code changed.

No more need for manual testing?

It's tempting to think that using TDD as described in this book might eliminate manual testing. It does eliminate some manual processes, but certainly not all. The main manual steps we replace with automation are feature testing during development and regression testing before release.

As we develop a new feature with TDD, we start by writing automated tests for that feature. Every automated test we write is a test that does not need to be run by hand. We save all that test setup time, together with the often lengthy process to click through a user interface to trigger the behavior we're testing. The main difference TDD brings is replacing test plans written in a word processor with test code written in an IDE. Development feature manual testing is replaced by automation.

TDD also provides us with automated regression testing, for free:

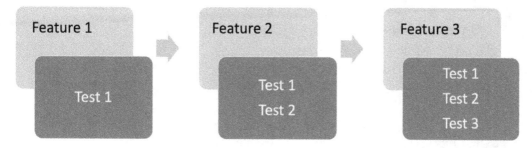

Figure 11.1 – Regression testing

Using TDD, we add one or more tests as we build out each feature. Significantly, we retain all those tests. We naturally build up a large suite of automated tests, captured in source control and executed on every build automatically. This is known as a regression test suite. Regression testing means that we re-check all the tests run to date on every build. This ensures that as we make changes to the system, we don't break anything. Moving fast and not breaking things might be how we describe this approach.

Regression tests also include tests for previously reported defects. These regression tests confirm that they have not been re-introduced. It bears repeating that the regression suite saves on all the manual effort required by non-automated tests *each and every* time the suite gets executed. This adds up over the full software life cycle to a huge reduction.

Test automation is good, but an automated test is a software machine. It cannot think for itself. It cannot visually inspect code. It cannot assess the appearance of a user interface. It cannot tell whether the user experience is good or bad. It cannot determine whether the overall system is fit for purpose.

This is where human-driven manual testing comes in. The following sections will look at areas where we need human-led testing, starting with an obvious one: finding bugs that our tests missed.

Manual exploratory – discovering the unexpected

In this section, we will appreciate the role of manual exploratory testing as an important line of defense against defects where TDD is used.

The biggest threat to our success with TDD lies in our ability to think about all the conditions our software needs to handle. Any reasonably complex piece of software has a huge range of possible input combinations, edge cases, and configuration options.

Consider using TDD to write code to restrict the sales of a product to buyers who are 18 years old and above. We must first write a happy-path test to check whether the sale is allowed, make it pass, then write a negative test, confirming that the sale can be blocked based on age. This test has the following form:

```java
public class RestrictedSalesTest {

    @Test
    void saleRestrictedTo17yearOld() {
        // ... test code omitted
    }


    @Test
    void salePermittedTo19yearOld() {
        // ... test code omitted
    }
}
```

The error is obvious when we're looking for it: what happens at the boundary between the ages of 17 and 18? Can an 18-year-old buy this product or not? We don't know, because there is no test for that. We tested for 17 and 19 years old. For that matter, what should happen on that boundary? In general, that's a stakeholder decision.

Automated tests cannot do two things:

- Ask a stakeholder what they want the software to do
- Spot a missing test

This is where manual exploratory testing comes in. This is an approach to testing that makes the most of human creativity. It uses our instincts and intelligence to work out what tests we might be missing. It then uses scientific experimentation to discover whether our predictions of a missing test were correct. If proven true, we can provide feedback on these findings and repair the defect. This can be done either as an informal discussion or using a formal defect tracking tool. In due course, we can write new automated tests to capture our discoveries and provide regression tests for the future.

This kind of exploratory testing is a highly technical job, based on knowledge of what kinds of boundaries exist in software systems. It also requires extensive knowledge of local deployment and setup of software systems, together with knowing how software is built, and where defects are likely to appear. To an extent, it relies on knowing how developers think and predicting the kinds of things they may overlook.

Some key differences between automated testing and exploratory testing can be summarized as follows:

Automated Testing	Manual Exploratory Testing
Repeatable	Creative
Tests for known outcomes	Finds unknown outcomes
Possible by machine	Requires human creativity
Behavior verification	Behavior investigation
Planned	Opportunistic
Code is in control of the testing	Human minds control the testing

Table 11.1 – Automated versus manual exploratory testing

Manual exploratory testing will always be needed. Even the best developers get pressed for time, distracted, or have yet another meeting that should have been an email. Once concentration is lost, it's all too easy for mistakes to creep in. Some missing tests relate to edge cases that we cannot see alone. Another human perspective often brings a fresh insight we would simply never have unaided. Manual exploratory testing provides an important extra layer of defense in depth against defects going unnoticed.

Once exploratory testing identifies some unexpected behavior, we can feed this back into development. At that point, we can use TDD to write a test for the correct behavior, confirm the presence of the defect, then develop the fix. We now have a fix and a regression test to ensure the bug remains fixed. We can think of manual exploratory testing as the fastest possible feedback loop for a defect we missed. An excellent guide to exploratory testing is listed in the *Further reading* section.

Seen in this light, automation testing and TDD do not make manual efforts less important. Instead, their value is amplified. The two approaches work together to build quality into the code base.

Manual testing for things we missed isn't the only development time activity of value that cannot be automated. We also have the task of checking the quality of our source code, which is the subject of the next section.

Code review and ensemble programming

This section reviews another area surprisingly resistant to automation: checking code quality.

As we've seen throughout this book, TDD is primarily concerned with the design of our code. As we build up a unit test, we define how our code will be used by its consumers. The implementation of that design is of no concern to our test, but it does concern us as software engineers. We want that implementation to perform well and to be easy for the next reader to understand. Code is read many more times than it is written over its life cycle.

Some automated tools exist to help with checking code quality. These are known as static code analysis tools. The name comes from the fact that they do not run code; instead, they perform an automated review of the source code. One popular tool for Java is Sonarqube (at `https://www.sonarqube.org/`), which runs a set of rules across a code base.

Out of the box, tools like this give warnings about the following:

- Variable name conventions not being followed
- Uninitialized variables leading to possible `NullPointerException` problems
- Security vulnerabilities
- Poor or risky use of programming constructs
- Violations of community-accepted practices and standards

These rules can be modified and added to, allowing customization to be made to the local project house style and rules.

Of course, such automated assessments have limitations. As with manual exploratory testing, there are simply some things only a human can do (at least at the time of writing). In terms of code analysis, this mainly involves bringing context to the decisions. One simple example here is preferring longer, more descriptive variable names to a primitive such as `int`, compared to a more detailed type such as `WordRepository`. Static tools lack that understanding of the different contexts.

Automated code analysis has its benefits and limitations, as summarized here:

Automated Analysis	Human Review
Rigid rules (for example, variable name length)	Relaxes rules based on context
Applies a fixed set of assessment criteria	Applies experiential learning
Reports pass/fail outcomes	Suggests alternative improvements

Table 11.2 – Automated analysis versus human review

Google has a very interesting system called **Google Tricorder**. This is a set of program analysis tools that combines the creativity of Google engineers in devising rules for good code with the automation to apply them. For more information, see `https://research.google/pubs/pub43322/`.

Manually reviewing code can be done in various ways, with some common approaches:

- **Code review on pull requests**:

  A pull request, also known as a merge request, is made by a developer when they wish to integrate their latest code changes into the main code base. This provides an opportunity for another developer to review that work and suggest improvements. They may even visually spot defects. Once the original developer makes agreed changes, the request is approved and the code is merged.

- **Pair programming**:

  Pair programming is a way of working where two developers work on the same task at the same time. There is a continuous discussion about how to write the code in the best way. It is a continuous review process. As soon as either developer spots a problem, or has a suggested improvement, a discussion happens and a decision is made. The code is continuously corrected and refined as it is developed.

- **Ensemble (mob) programming**:

  Like pair programming, only the whole team takes part in writing the code for one task. The ultimate in collaboration, which continuously brings the expertise and opinions of an entire team to bear on every piece of code written.

The dramatic difference here is that a code review happens after the code is written, but pair programming and mobbing happen while the code is being written. Code reviews performed after the code is written frequently happen too late to allow meaningful changes to be made. Pairing and mobbing avoid this by reviewing and refining code continuously. Changes are made the instant they are identified. This can result in higher quality output delivered sooner compared to the code-then-review workflow.

Different development situations will adopt different practices. In every case, adding a second pair of human eyes (or more) provides an opportunity for a design-level improvement, not a syntax-level one.

With that, we've seen how developers can benefit from adding manual exploratory testing and code review to their TDD work. Manual techniques benefit our users as well, as we will cover in the next section.

User interface and user experience testing

In this section, we will consider how we evaluate the impact of our user interface on users. This is another area where automation brings benefits but cannot complete the job without humans being involved.

Testing the user interface

User interfaces are the only part of our software system that matters to the most important people of all: our users. They are – quite literally – their windows into our world. Whether we have a command-line interface, a mobile web application, or a desktop GUI, our users will be helped or hindered in their tasks by our user interface.

The success of a user interface rests on two things being done well:

- It provides all the functionality a user needs (and wants)
- It allows a user to accomplish their end goals in an effective and efficient manner

The first of these, providing functionality, is the more programmatic of the two. In the same way that we use TDD to drive a good design for our server-side code, we can use it in our frontend code as well. If our Java application generates HTML – called server-side rendering – TDD is trivial to use. We test the

HTML generation adapter and we're done. If we are using a JavaScript/Typescript framework running in the browser, we can use TDD there, with a test framework such as Jest (`https://jestjs.io/`).

Having tested we're providing the right functions to the user, automation then becomes less useful. With TDD, we can verify that all the right sorts of graphical elements are present in our user interface. But we can't tell whether they are meeting the needs of the user.

Consider this fictional user interface for buying merchandise relating to our Wordz application:

Buy Products

Product name:

Wordz T-Shirt (M)

Quantity:

1

Submit

Figure 11.2 – Example user interface

We can use TDD to test that all those interface elements – the boxes and buttons – are present and working. But will our users care? Here are the questions we need to ask:

- Does it look and feel good?
- Does it align with corporate branding and house style guides?
- For the task of buying a T-shirt, is it easy to use?
- Does it present a logical flow to the user, guiding them through their task?

Quite deliberately for this example, the answer is no to all these questions. This is, quite frankly, a terrible user interface layout. It has no style, no feeling, and no brand identity. You have to type in the product name in the text field. There is no product image, no description, and no price! This user interface is truly the worst imaginable for an e-commerce product sales page. Yet it would pass all our automated functionality tests.

Designing effective user interfaces is a very human skill. It involves a little psychology in knowing how humans behave when presented with a task, mixed with an artistic eye, backed by creativity. These qualities of a user interface are best assessed by humans, adding another manual step to our development process.

Evaluating the user experience

Closely related to user interface design is user experience design.

User experience goes beyond any individual element or view on a user interface. It is the entire experience our users have, end to end. When we want to order the latest Wordz T-shirt from our e-commerce store, we want the entire process to be easy. We want the workflow across every screen to be obvious, uncluttered, and easier to get right than to get wrong. Going further, service design is about optimizing the experience from wanting a T-shirt to wearing it.

Ensuring users have a great experience is the job of a user experience designer. It is a human activity that combines empathy, psychology, and experimentation. Automation is limited in how it can help here. Some mechanical parts of this can be automated. Obvious candidates are applications such as Invision (`https://www.invisionapp.com/`), which allows us to produce a screen mockup that can be interacted with, and Google Forms, which allows us to collect feedback over the web, with no code to set that up.

After creating a candidate user experience, we can craft experiments where potential users are given a task to complete, then asked to provide feedback on how they found the experience.

A simple, manual form is more than adequate to capture this feedback:

Experience	Rating of 1 (Poor) – 5 (Good)	Comments
My task was easy to complete	4	I completed the task ok after being prompted by your researcher.
I felt confident completing my task without instructions	2	The text entry field about T-shirt size confused me. Could it be a dropdown of available options instead?
The interface guided me through the task	3	It was ok in the end – but that text field was an annoyance, so I scored this task lower.

Table 11.3 – User experience feedback form

User experience design is primarily a human activity. So is the evaluation of test results. These tools only go as far as allowing us to create a mockup of our visions and collect experimental results. We must run sessions with real users, solicit their opinions on how their experience was, then feed back the results in an improved design.

While user experience is important, the next section deals with a mission-critical aspect of our code: security and operations.

Security testing and operations monitoring

This section reflects on the critical aspects of security and operations concerns.

So far, we have created an application that is well-engineered and has very low defects. Our user experience feedback has been positive – it is easy to use. But all that potential can be lost in an instant if we cannot keep the application running. If hackers target our site and harm users, the situation becomes even worse.

An application that is not running does not exist. The discipline of operations – often called DevOps these days – aims to keep applications running in good health and alert us if that health starts to fail.

Security testing – also called **penetration testing** (**pentesting**) – is a special case of manual exploratory testing. By its nature, we are looking for new exploits and unknown vulnerabilities in our application. Such work is not best served by automation. Automation repeats what is already known; to discover the unknown requires human ingenuity.

Penetration testing is the discipline that takes a piece of software and attempts to circumvent its security. Security breaches can be expensive, embarrassing, or business-ending for a company. The exploits used to create the breach are often very simple.

Security risks can be summarized roughly as follows:

- Things we shouldn't see
- Things we shouldn't change
- Things we shouldn't use as often
- Things we should not be able to lie about

This is an oversimplification, of course. But the fact remains that our application may be vulnerable to these damaging activities – and we need to know whether that is the case or not. This requires testing. This kind of testing must be adaptive, creative, devious, and continually updated. An automated approach is none of those things, meaning security testing must take its place as a manual step in our development process.

A great starting point is to review the latest **OWASP Top 10 Web Application Security Risks** (`https://owasp.org/www-project-top-ten/`) and begin some manual exploratory testing based on the risks listed. Further information on threat models such as **Spoofing, Tampering, Repudiation, Information Disclosure, Denial of Service, and Elevation of Privilege** (STRIDE) can be found at `https://www.eccouncil.org/threat-modeling/`. OWASP also has some excellent resources on useful tools at `https://owasp.org/www-community/Fuzzing`. **Fuzzing** is an automated way of discovering defects, although it requires a human to interpret the results of a failed test.

As with other manual exploratory tests, these ad hoc experiments may lead to some future test automation. But the real value lies in the creativity applied to investigating the unknown.

The preceding sections have made a case for the importance of manual interventions to complement our test automation efforts. But how does that fit in with a **continuous integration/continuous delivery (CI/CD)** approach? That's the focus of the next section.

Incorporating manual elements into CI/CD workflows

We've seen that not only are manual processes important in our overall workflow but for some things, they are irreplaceable. But how do manual steps fit into heavily automated workflows? That's the challenge we will cover in this section.

Integrating manual processes into an automated CI/CD pipeline can be difficult. The two approaches are not natural partners in terms of a linear, repeatable sequence of activities. The approach we take depends on our ultimate goal. Do we want a fully automated continuous deployment system, or are we happy with some manual interruptions?

The simplest approach to incorporating a manual process is to simply stop the automation at a suitable point, begin the manual process, then resume automaton once the manual process completes. We can think of this as a blocking workflow, as all further automated steps in the pipeline must stop until the manual work is completed. This is illustrated in the following diagram:

Figure 11.3 – Blocking workflow

By organizing our development process as a series of stages, some being automated and some being manual, we create a simple blocking workflow. Blocking here means that the flow of value is blocked by each stage. The automation stages typically run more quickly than the manual stages.

This workflow has some advantages in that it's simple to understand and operate. Each iteration of software we deliver will have all automated tests run as well as all the current manual processes. In one sense, this release is of the highest quality we know how to make at that time. The disadvantage is that each iteration must wait for all manual processes to complete:

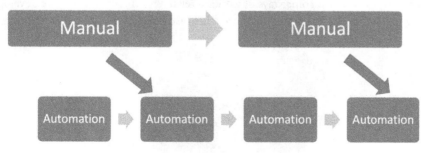

Figure 11.4 – Dual track workflow

One enabler for very smooth dual-track workflows is to use a single main trunk for the whole code base. All developers commit to this main trunk. There are no other branches. Any features in active development are isolated by **feature flags**. These are Boolean values that can be set to `true or false` at runtime. The code inspects these flags and decides whether to run a feature or not. Manual testing can then happen without having to pause deployments. During testing, the features in progress are enabled via the relevant feature flags. For the general end users, features in progress are disabled.

We can select the approach that fits our delivery goals the best. The blocking workflow trades off less rework for an extended delivery cycle. The dual-track approach allows for more frequent feature delivery, with a risk of having defects in production before they are discovered by a manual process and, subsequently, repaired.

Selecting the right process to use involves a trade-off between feature release cadence and tolerating defects. Whatever we choose, the goal is to focus the expertise of the whole team on creating software with a low defect rate.

Balancing automated workflows with manual, human workflows isn't easy, but it does result in getting the most human intuition and experience into the product. That's good for our development teams and it is good for our users. They benefit from improved ease of use and robustness in their applications. Hopefully, this chapter has shown you how we can combine these two worlds and cross that traditional developer-tester divide. We can make one great team, aiming at one excellent outcome.

Summary

This chapter discussed the importance of various manual processes during development.

Despite its advantages, we've seen how TDD cannot prevent all kinds of defects in software. First, we covered the benefits of applying human creativity to manual exploratory testing, where we can uncover defects that we missed during TDD. Then, we highlighted the quality improvements that code reviews and analysis bring. We also covered the very manual nature of creating and verifying excellent user interfaces with satisfying user experiences. Next, we emphasized the importance of security testing and operations monitoring in keeping a live system working well. Finally, we reviewed approaches to integrating manual steps into automation workflows, and the trade-offs we need to make.

In the next chapter, we'll review some ways of working related to when and where we develop tests, before moving on to *Part 3* of this book, where we will finish building our Wordz application.

Questions and answers

The following are some questions and answers regarding this chapter's content:

1. Have TDD and CI/CD pipelines eliminated the need for manual testing?

    No. They have changed where the value lies. Some manual processes have become irrelevant, whereas others have increased in importance. Traditionally, manual steps, such as following test documents for feature testing and regression testing, are no longer required. Running feature and regression tests has changed from writing test plans in a word processor to writing test code in an IDE. But for many human-centric tasks, having a human mind in the loop remains vital to success.

2. Will **artificial intelligence (AI)** automate away the remaining tasks?

    This is unknown. Advances in AI at this time (the early 2020s) can probably improve visual identification and static code analysis. It is conceivable that AI image analysis may one day be able to provide a good/bad analysis of usability – but that is pure speculation, based on AI's abilities to generate artworks today. Such a thing may remain impossible. In terms of practical advice now, assume that the recommended manual processes in this chapter will remain manual for some time.

Further reading

To learn more about the topics that were covered in this chapter, take a look at the following resources:

*   `https://dl.acm.org/doi/pdf/10.1145/274567.274574`:

    An overview of the modern genesis of TDD by Kent Beck. While the ideas certainly predate this project, this is the central reference of modern TDD practice. This paper contains many important insights into software development and teams – including the quote make it run, make it right, make it fast, and the need to not feel like we are working all the time. Well worth reading.

*   Explore It, Elizabeth Hendrickson, ISBN 978-1937785024.

*   `https://trunkbaseddevelopment.com/`.

*   `https://martinfowler.com/articles/feature-toggles.html`.

*   Inspired: How to create tech products customers love, Marty Cagan, ISBN 978-1119387503:

    An interesting book that talks about product management. While this may seem strange in a developer book on TDD, a lot of the ideas in this chapter came from developer experience in a dual-track agile project, following this book. Dual agile means that fast feedback loops on feature discovery feed into fast feedback agile/TDD delivery. Essentially, manual TDD is done at the product requirements level. This book is an interesting read regarding modern product management, which has adopted the principles of TDD for rapid validation of assumptions about user features. Many ideas in this chapter aim to improve the software at the product level.

12

Test First, Test Later, Test Never

In this chapter, we are going to review some of the nuances of **Test-Driven Development** (TDD). We've already covered the broad techniques of writing unit tests as part of an overall test strategy. We can use the test pyramid and hexagonal architecture to guide the scope of our tests in terms of what specifically they need to cover.

We have two more dimensions we need to decide on: when and where to start testing. The first question is one of timing. Should we always write our tests before the code? What difference would it make to write tests after the code? In fact, what about not testing at all – does that ever make sense? Where to start testing is another variable to decide. There are two schools of thought when it comes to TDD – testing from the inside out or the outside in. We will review what these terms mean and what impact each has on our work. Finally, we will consider how these approaches work with a hexagonal architecture to form a natural testing boundary.

In this chapter we're going to cover the following main topics:

- Adding tests first
- We can always test it later, right?
- Tests? They're for people who can't write code!
- Testing from the inside out
- Testing from the outside in
- Defining test boundaries with hexagonal architecture

Adding tests first

In this section, we will review the trade-offs of adding a test first before writing the production code to make it pass.

Previous chapters have followed a test-first approach to writing code. We write a test before writing production code to make that test pass. This is a recommended approach, but it is important to understand some of the difficulties associated with it as well as considering its benefits.

Test-first is a design tool

The most important benefit of writing tests first is that a test acts as a *design aid*. As we decide what to write in our test, we are designing the interface to our code. Each of the test stages helps us consider an aspect of software design, as illustrated by the following diagram:

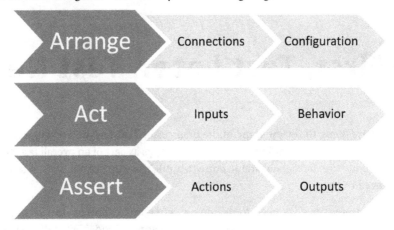

Figure 12.1 – Test-first aids design

The **Arrange** step helps us think about how the code under test relates to the bigger picture of the whole code base. This step helps us design how the code will fit into the whole code base. It gives us an opportunity to make the following design decisions:

- What configuration data will be needed?
- What connections to other objects or functions will be needed?
- What behavior should this code provide?
- What extra inputs are needed to provide that behavior?

Coding the Act step allows us to think about how easy our code will be to use. We reflect on what we would like the method signature of the code we are designing to be. Ideally, it should be simple and unambiguous. Some general recommendations are the following:

- The method name should describe the outcome of calling the method.
- Pass in as few parameters as possible. Possibly group parameters into their own object.
- Avoid Boolean flags that modify the behavior of the code. Use separate methods with appropriate names.
- Avoid requiring multiple method calls to do one thing. It is too easy to miss out on an important call in the sequence if we are unfamiliar with the code.

Writing the Act step allows us to see what the call to our code will look like everywhere it is used for the first time. This provides the opportunity to simplify and clarify before our code gets widely used.

The code in our Assert step is the first consumer of the results of our code. We can judge from this step whether those results are easy to obtain. If we are unhappy with how the Assert code looks, this is a chance to review how our object provides its output.

Every test we write provides this opportunity for a design review. TDD is all about helping us uncover better designs, even more than it is about testing correctness.

In other industries, such as designing cars, it is common to have dedicated design tools. **AutoCAD 3D Studio** is used to create 3D models for the chassis of a car on a computer. Before we manufacture the car, we can use the tool to pre-visualize the end result, rotating it through space and viewing it from several angles.

Mainstream commercial software engineering lags far behind in terms of design tool support. We don't have an equivalent to 3D Studio for designing code. The 1980s to 2000s saw the rise of **Computer-Aided Software Engineering** (CASE) **tools** but these appear to have fallen into disuse. CASE tools purported to simplify software engineering by allowing their users to enter various graphical forms of software structures, then generate code that implemented those structures. Today, writing TDD tests prior to writing the production code seems to be the closest thing we have to computer-aided design for software at present.

Tests form executable specifications

Another advantage of test code is that it can form a highly accurate, repeatable form of documentation. Simplicity and clarity in the test code are required to achieve that. Instead of writing a test planning document, we write TDD tests as code, which can be run by a computer. This has the benefit of being more immediate for developers. These executable specifications are captured alongside the production code they test, stored in source control, and made continuously available to the whole team.

Further documentation is useful. Documents such as **RAID logs** – documenting risks, actions, issues, and decisions – and **KDDs** – documenting key design decisions – are often required. These are non-executable documents. They serve the purpose of capturing who, when, and critically *why* an important decision was made. Information of this kind cannot be captured using test code, meaning that these kinds of documents have value.

Test-first provides meaningful code coverage metrics

Writing a test before we write production code gives each test a specific purpose. The test exists to drive out a specific behavior in our code. Once we get this test to pass, we can run the test suite using a code coverage tool, which will output a report similar to the following:

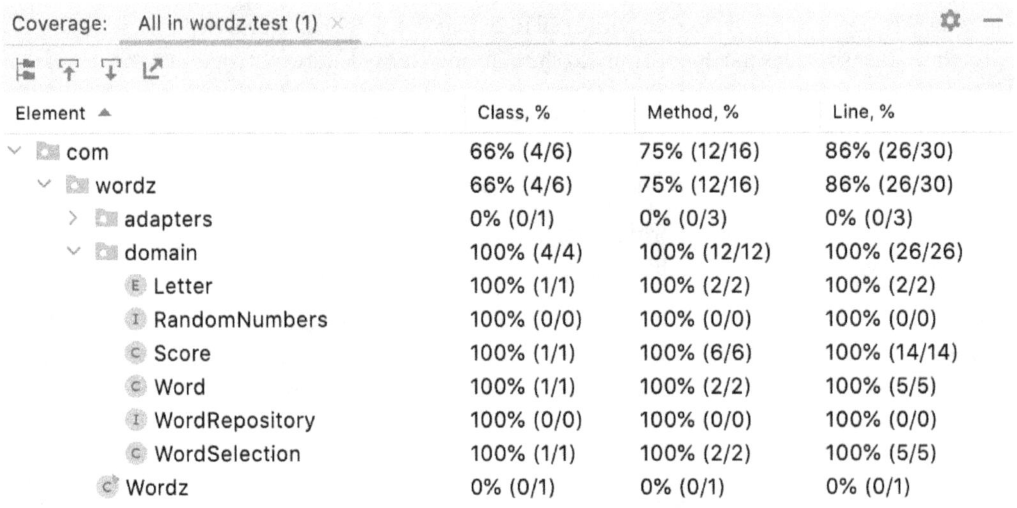

Coverage: All in wordz.test (1) ×			⚙ —
⊫ ⬆ ⬇ ↗			
Element ▲	Class, %	Method, %	Line, %
∨ ▣ com	66% (4/6)	75% (12/16)	86% (26/30)
∨ ▣ wordz	66% (4/6)	75% (12/16)	86% (26/30)
› ▣ adapters	0% (0/1)	0% (0/3)	0% (0/3)
∨ ▣ domain	100% (4/4)	100% (12/12)	100% (26/26)
⒠ Letter	100% (1/1)	100% (2/2)	100% (2/2)
⒤ RandomNumbers	100% (0/0)	100% (0/0)	100% (0/0)
⒞ Score	100% (1/1)	100% (6/6)	100% (14/14)
⒞ Word	100% (1/1)	100% (2/2)	100% (5/5)
⒤ WordRepository	100% (0/0)	100% (0/0)	100% (0/0)
⒞ WordSelection	100% (1/1)	100% (2/2)	100% (5/5)
⒞ Wordz	0% (0/1)	0% (0/1)	0% (0/1)

Figure 12.2 – Code coverage report

A **code coverage** tool instruments our production code as we run the tests. This instrumentation captures which lines of code were executed during running the tests. This report can suggest we have missing tests, by flagging up lines of code that were never executed during the test run.

The code coverage report in the image shows we have executed 100% of the code in the domain model by our test run. Having 100% coverage is entirely down to us writing a TDD test before we write code to make it pass. We do not add untested code with a test-first TDD workflow.

Beware of making a code coverage metric a target

A high code coverage metric doesn't always indicate high code quality. If we are writing tests for generated code or tests for code we've pulled from a library, that coverage does not tell us anything new. We may assume – generally – that our code generators and libraries have already been tested by their developers.

However, a real problem with code coverage numbers happens when we mandate them as a metric. As soon as we impose a minimum coverage target on developers, then **Goodhart's law** applies – *when a measure becomes a target, it ceases to be a good measure.* Humans will sometimes cheat the system to achieve a target when under pressure. When that happens, you see code such as this:

```
public class WordTest {
    @Test
    public void oneCorrectLetter() {
        var word = new Word("A");
        var score = word.guess("A");
```

```
        // assertThat(score).isEqualTo(CORRECT);
    }
}
```

Notice those comment symbols – // – just before `assertThat()`? That's the hallmark of a test case that was failing and could not be made to pass by a certain deadline. By retaining the test, we keep our number of test cases up, and we keep our code coverage percentage up. A test such as this will execute lines of production code, but it will not *validate* that they work. The code coverage target will be hit – even though the code itself does not work.

Now, I know what you're thinking – no developer would ever cheat the test code like this. It is, however, an example from a project I worked on for a major international client. The client had engaged both the company I work for and another development team to work on some microservices. Due to a time zone difference, the other team would check in their code changes while our team was asleep.

We came in one morning to see our test results dashboards lit up red. The overnight code change had caused large numbers of our tests to fail. We checked the pipelines of the other team and were astonished to see all their tests passing. This made no sense. Our tests clearly revealed a defect in that nightly code drop. We could even localize it from our test failures. This defect would have shown up in the unit tests around that code, but those unit tests were passing. The reason? Commented-out asserts.

The other team was under pressure to deliver. They obeyed their instructions to get that code change checked in on that day. Those changes, in fact, had broken their unit tests. When they could not fix them in the time available, they chose to cheat the system and defer the problem to another day. I'm not sure I blame them. Sometimes, 100% code coverage and all tests passing mean nothing at all.

Beware of writing all tests upfront

One of the strengths of TDD is that it allows for *emergent design*. We do a small piece of design work, captured in a test. We then do the next small piece of design, captured in a new test. We perform varying depths of refactoring as we go. In this way, we learn about what is and is not working in our approach. The tests provide fast feedback on our design.

This can only happen if we write tests one at a time. A temptation for those familiar with waterfall project approaches can be to treat the test code as one giant requirements document, to be completed before development starts. While this seems more promising than simply writing a requirements document in a word processor, it also means that developers cannot learn from test feedback. There is no feedback cycle. This approach to testing should be avoided. Better results are obtained by taking an incremental approach. We write one test at a time, together with the production code to make that test pass.

Writing tests first helps with continuous delivery

Perhaps the biggest benefit of writing tests first lies in continuous delivery situations. Continuous delivery relies on a highly automated pipeline. Once a code change is pushed to source control, the build pipeline is started, all tests run, and finally, a deployment occurs.

The only reason for code not to deploy in this system – assuming the code compiles – is if the tests fail. This implies that the automated tests we have in place are necessary and sufficient to create the level of confidence required.

Writing tests first cannot guarantee this – we may still have missing tests – but out of all the ways of working with tests, it is perhaps the most likely to result in one meaningful test for each piece of application behavior that we care about.

This section has presented the case that writing tests first – before production code is written, to make them pass – helps create confidence in our code, as well as useful executable specifications. However, that's not the only way to code. Indeed, a common approach we will see is to write a chunk of code first and then write tests shortly after.

The next section looks at the advantages and limitations of the test-later approach.

We can always test it later, right?

An alternative approach to writing tests before code is to write code first, then write tests. This section compares and contrasts writing tests after the code with writing tests before the code.

One approach to writing tests involves writing chunks of code and then retrofitting tests to those pieces of code. It's an approach that is used in commercial programming, and the workflow can be illustrated as follows:

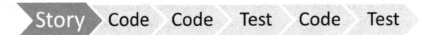

Figure 12.3 – Test-after workflow

Upon selecting a user story to develop, one or more pieces of production code are written. Tests follow! Academic research seems mixed, to say the least, on whether or not test-after differs from test-first. From one 2014 study by the ACM, an extract from the conclusion was this:

> *"...static code analysis results were found statistically significant in the favor of TDD. Moreover, the results of the survey revealed that the majority of developers in the experiment prefer TLD over TDD, given the lesser required level of learning curve."*

(Source: https://dl.acm.org/doi/10.1145/2601248.2601267)

However, a commenter pointed out that in this research, the following applied:

> *"...usable data was obtained from only 13 out of 31 developers. This meant the statistical analysis was undertaken using groups of seven (TDD) and six (TLD). There is no real surprise that the experiment was found to lack statistical power and that the findings were inconclusive."*

Other research papers seem to show similar lackluster results. Practically then, what should we take away from this? Let's consider some practical details of test-later development.

Test-later is easier for a beginner to TDD

One finding of the research was that beginners to TDD found test-later to be easier to get started with. This seems reasonable. Before we attempt TDD, we may consider coding and testing as different activities. We write code according to some set of heuristics, and then we figure out how to test that code. Adopting a test-later approach means that the coding phase is essentially unchanged by the demands of testing. We can continue coding as we always did. There is no impact from having to consider the impacts of testing on the design of that code. This seeming advantage is short-lived, as we discover the need to add access points for testing, but we can at least get started easily.

Adding tests later works reasonably well if we keep writing tests in lockstep with the production code: write a little code, and write a few tests for that code – but not having tests for every code path remains a risk.

Test-later makes it harder to test every code path

A plausible argument against using a test-later approach is that it becomes harder to keep track of having all the tests we need. On the face of it, this claim cannot be completely true. We can always find some way to keep track of the tests we need. A test is a test, no matter when it is written.

The problem comes as the time between adding tests increases. We are adding more code, which means adding more execution paths throughout the code. For example, every `if` statement we write represents two execution paths. Ideally, every execution path through our code will have a test. Every untested execution path we add places us one test below this ideal number. This is illustrated directly in flowcharts:

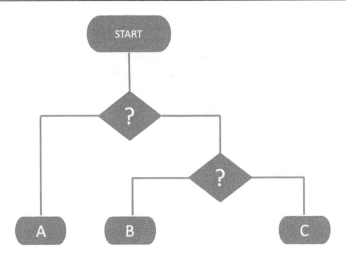

Figure 12.4 – Illustrating execution paths

This flowchart depicts a process with nested decision points – the diamond shapes – which result in three possible execution paths, labeled **A**, **B**, and **C**. The technical measure of the number of execution paths is called **cyclomatic complexity**. The complexity score is the number calculated on how many linearly independent execution paths exist in a piece of code. The code in the flowchart has a cyclomatic complexity of three.

As we increase the cyclomatic complexity of our code, we increase our *cognitive load* with the need to remember all those tests that we need to write later. At some point, we might even find ourselves periodically stopping coding and writing down notes for what tests to add later. This sounds like a more arduous version of simply writing the tests as we go.

The issue of keeping track of tests we are yet to write is avoided when using test-first development.

Test-later makes it harder to influence the software design

One of the benefits of test-first development is that the feedback loop is very short. We write one test and then complete a small amount of production code. We then refactor as required. This moves away from a waterfall-style pre-planned design to an emergent design. We change our design in response to learning more about the problem we are solving as we incrementally solve more of it.

When writing tests after a chunk of code has already been written, it gets harder to incorporate feedback. We may find that the code we have created proves difficult to integrate into the rest of the code base. Perhaps this code is confusing to use due to having unclear interfaces. Given all the effort we have spent creating the messy code, it can be tempting to just live with the awkward design and its equally awkward test code.

Test-later may never happen

Development tends to be a busy activity, especially when deadlines are involved. Time pressures may mean that the time we hoped to get to write our tests simply never comes. It's not uncommon for project managers to be more impressed with new features than with tests. This seems a false economy – as users only care about features that *work* – but it's a pressure that developers sometimes face.

This section has shown that writing tests shortly after writing code can work as well as writing tests first if care is exercised. It also seems preferable to some developers at the start of their TDD journey – but what about the ultimate extreme of *never* testing our code? Let's quickly review the consequences of that approach.

Tests? They're for people who can't write code!

This section discusses another obvious possibility when it comes to automated testing – simply not writing automated tests at all. Perhaps not even testing at all. Is this viable?

Not testing at all is a choice we could make, and this might not be as silly as it sounds. If we define testing as *verifying some outcome is achieved in its target environment,* then things such as deep-space probes cannot truly be tested on Earth. At best, we are simulating the target environment during our testing. Giant-scale web applications can rarely be tested with realistic load profiles. Take any large web application, launch a hundred million users at it – all doing invalid things – and see how most applications hold up. It's probably not as well as developer testing suggested.

There are areas of development where we might expect to see fewer automated tests:

- **Extract, Transform, and Load (ETL) scripts for data migrations**:

  ETL scripts are often one-off affairs, written to solve a specific migration problem with some data. It's not always worth writing automated tests for these, performing manual verification on a similar set of source data instead.

- **Front-end user interface work**:

  Depending on the programming approach, it may be challenging to write unit tests for the frontend code. Whatever approach we take, assessing the visual look and feel cannot currently be automated. As a result, manual testing is often used against a candidate release of a user interface.

- **Infrastructure-as-code scripts**:

  Our applications need to be deployed somewhere for them to run. A recent approach to deployment is to use languages such as Terraform to configure servers using code. This is an area that's not yet simple to automate tests for.

So what actually happens when we abandon test automation, possibly not even testing at all?

What happens if we do not test during development?

We might think that not testing at all is an option, but in reality, testing will always happen at some point. We can illustrate this with a timeline of the possible points at which testing *can* occur:

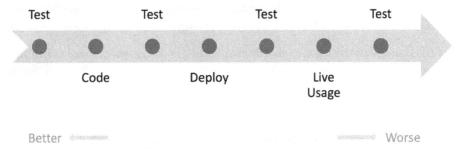

Figure 12.5 – Testing timeline

Test-first approaches shift the testing to be as early as possible – an approach called **shift-left** – where defects can be corrected cheaply and easily. Thinking that we won't test merely pushes testing all the way to the right – after users start using features live.

Ultimately, *all code that users care about gets tested eventually*. Maybe developers don't test it. Maybe testing will fall to another specialist testing team, who will write defect reports. Maybe defects will be found during the operation of the software. Most commonly of all, we end up outsourcing testing to the users themselves.

Having users test our code for us is generally a bad idea. Users trust us to give them software that solves their problems. Whenever a defect in our code prevents that from happening, we lose that trust. A loss of trust damages the 3 Rs of a business: revenue, reputation, and retention. Users may well switch to another supplier, whose better-tested code actually solves the user's problem.

If there is any possibility at all to test our work before we ship it, we should take that opportunity. The sooner we build test-driven feedback loops into our work, the easier it will be to improve the quality of that work.

Having looked at *when* we test our software, let's turn to *where* we test it. Given the overall design of a piece of software, where should we start testing? The next section reviews a test approach that starts from the inside of a design and works its way out.

Testing from the inside out

In this section, we're going to review our choice of starting point for our TDD activities. The first place to look at is inside our software system, starting with details.

When starting to build software, we obviously need some place to start from. One place to start is with some of the details. Software is made up of small interconnecting components, each of which performs a portion of the whole task. Some components come from library code. Many components are custom-made to provide the functionality our application needs.

One place to start building, then, is on the *inside* of this software system. Starting with an overall user story, we can imagine a small component that is likely to be of use to us. We can begin our TDD efforts around this component and see where that leads us. This is a bottom-up approach to the design, composing the whole from smaller parts.

If we consider a simplified version of our Wordz application structure, we can illustrate the inside-out approach as follows:

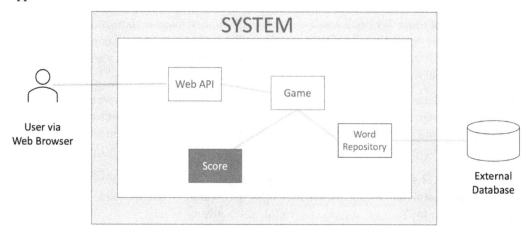

Figure 12.6 – Inside-out development

The diagram shows the **Score** component highlighted, as that is where we will start development using an inside-out approach. The other software components are grayed-out. We are not designing those pieces yet. We would start with a test for some behavior we wanted the **Score** component to have. We would work our way outward from that starting point.

This style of inside-out TDD is also known as **Classicist TDD** or **Chicago TDD**. It is the approach originally described by Kent Beck in his book *Test-Driven Development by Example*. The basic idea is to start anywhere to create any useful building block for our code. We then develop a progressively larger unit that incorporates the earlier building blocks.

The inside-out approach has a few advantages:

- **Quick start to development**: We test pure Java code first in this approach, using the familiar tools of JUnit and AssertJ. There is no setup for user interfaces, web service stubs, or databases. There is no setup of user interface testing tools. We just dive right in and code using Java.

- **Good for known designs**: As we gain experience, we recognize some problems as having known solutions. Perhaps we have written something similar before. Maybe we know a useful collection of design patterns that will work. In these cases, starting from the interior structure of our code makes sense.

- **Works well with hexagonal architecture**: Inside-out TDD starts work inside the inner hexagon, the domain model of our application. The adapter layer forms a natural boundary. An inside-out design is a good fit for this design approach.

Naturally, nothing is perfect and inside-out TDD is no exception. Some challenges include the following:

- **Possibility of waste**: We begin inside-out TDD with our best guess of some components that will be needed. Sometimes, it emerges later that either we don't need these components, or we should refactor the features somewhere else. Our initial effort is in some sense wasted – although it will have helped us progress to this point.

- **Risk of implementation lock-in**: Related to the previous point, sometimes we move on from an initial design having learned more about the problem we're solving, but we don't always recognize a sunk cost. There is always a temptation to keep using a component we wrote earlier even if it no longer fits as well, just because we invested that time and money into creating it.

Inside-out TDD is a useful approach and was first popularized by Kent Beck's book. However, if we can start inside-out, what about turning that around? What if we started from the outside of the system and worked our way in? The next section reviews this alternative approach.

Testing from the outside in

Given that inside-out TDD has some challenges as well as strengths, what difference does outside-in TDD make? This section reviews the alternative approach of starting from outside the system.

Outside-in TDD begins with the external users of the system. They may be human users or machines, consuming some API offered by our software. This approach to TDD begins by simulating some external input, such as the submission of a web form.

The test will typically use some kind of test framework – such as Selenium or Cypress for web applications – that allows the test to call up a specific web view, and simulate typing text into fields, then clicking a submit button. We can then make this test pass in the normal way, only we will have written some code that directly deals with the input from a user this time. In our hexagonal architecture model, we will end up writing the user input adapter first.

We can illustrate the outside-in approach as follows:

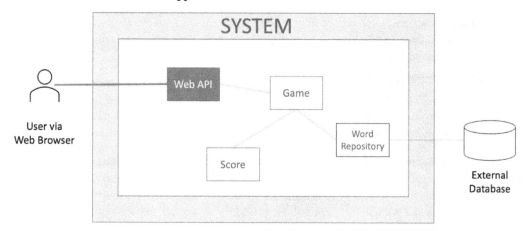

Figure 12.7 – Outside-in view

We can see that a component called **Web API** is the focus of our attention here. We will write a test that sets up enough of our application to run a component that handles web requests. The test will form a web request, send it to our software, and then assert that the correct web response is sent. The test may also instrument the software itself to verify it takes the expected actions internally. We start testing from the outside, and as development progresses, we move inwards.

This approach to TDD is described in the book, *Growing Object-Oriented Software, Guided by Tests*, by Steve Freeman and Nat Pryce. The technique is also known as the **London** or **Mockist** school of TDD. The reasons for that are the location where it was first popularized and its use of mock objects, respectively. To test drive the user input adapter as the first component we tackle, we need a test double in place of the rest of the software. Mocks and stubs are an inherent part of outside-in TDD.

Outside-in TDD, predictably enough, has some strengths and weaknesses. Let's take a look at the strengths first:

- **Less waste:** Outside-in TDD encourages a quite minimal approach to satisfying external behavior. The code produced tends to be highly customized to the application at hand. In contrast, inside-out TDD focuses on building a robust domain model, perhaps providing more functionality than will end up in use by users.

- **Delivers user value quickly:** Because we start from a test that simulates a user request, the code we write will satisfy a user request. We can deliver value to users almost immediately.

Outside-in TDD also has some weaknesses, or at least limitations:

- **Fewest abstractions**: On a related note, when writing the minimum code necessary to make a test pass, outside-in TDD may lead to application logic being present in the adapter layer. This can be refactored later but can lead to a less organized code base.

- **Inverted test pyramid**: If all our TDD test efforts focus on the external responses, they are, in fact, end-to-end tests. This opposes the recommended pattern of the test pyramid, which prefers faster unit tests inside the code base. Having only slower, less repeatable end-to-end tests can slow development.

The two traditional schools of TDD both offer certain advantages in terms of how they affect the software design we will produce. The next section looks at the impact of hexagonal architecture. By starting from the idea that we will use a hexagonal approach, we can combine the advantages of both schools of TDD. We end up defining a natural test boundary between the inside-out and outside-in approaches to TDD.

Defining test boundaries with hexagonal architecture

The topic for this section is how using a hexagonal architecture impacts TDD. Knowing that we are using hexagonal architecture presents useful boundaries for the different kinds of tests in the test pyramid.

In one sense, how we organize our code base does not affect our use of TDD. The internal structure of the code is simply an implementation detail, one of many possibilities that will make our tests pass. That being said, some ways of structuring our code are easier to work with than others. Using hexagonal architecture as a foundational structure does offer TDD some advantages. The reason why lies with the use of ports and adapters.

We've learned from previous chapters that it is easier to write tests for code where we can control the environment in which the code runs. We've seen how the test pyramid gives a structure to the different kinds of tests we write. Using the ports and adapters approach provides clean boundaries for each kind of test in the code. Better yet, it provides us with an opportunity to bring even more tests to the unit test level.

Let's review what kinds of tests best fit each layer of software written using hexagonal architecture.

Inside-out works well with the domain model

Classic TDD uses an inside-out development approach, where we choose a certain software component to test-drive. This component may be a single function, a single class, or a small cluster of classes that collaborate with each other. We use TDD to test this component as a whole given the behaviors it offers to its consumers.

This kind of component resides in the domain model – the inner hexagon:

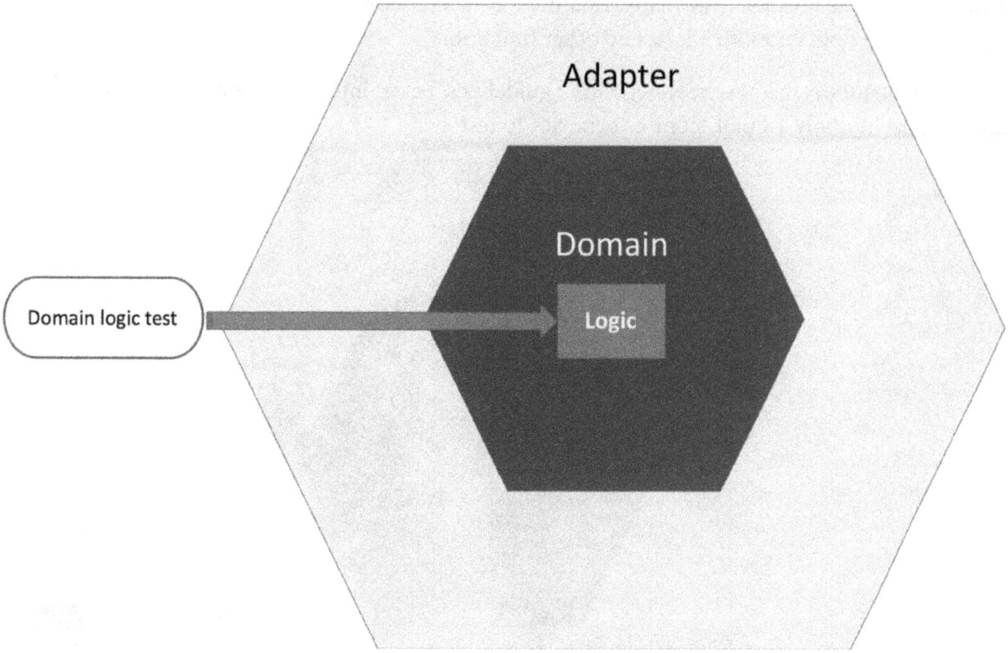

Figure 12.8 – Testing the domain logic

The key advantage is that these components are easy to write tests for and those tests run very quickly. Everything lives in computer memory and there are no external systems to contend with.

A further advantage is that complex behaviors can be unit-tested here at a very fine granularity. An example would be testing all the state transitions within a finite state machine used to control a workflow.

One disadvantage is that these fine-grained domain logic tests can get lost if a larger refactoring takes place. If the component under fine-grained tests gets removed during refactoring, its corresponding test will be lost – but the behavior will still exist somewhere else as a result of that refactoring. One thing refactoring tools cannot do is figure out what test code relates to the production code being refactored, and automatically refactor the test code to fit the new structure.

Outside-in works well with adapters

Mockist-style TDD approaches development from an outside-in perspective. This is a great match for our adapter layer in a hexagonal architecture. We can assume that the core application logic resides in the domain model and has been tested there with fast unit tests. This leaves adapters in the outer hexagon to be tested by integration tests.

These integration tests only need to cover the behavior provided by the adapter. This should be very limited in scope. The adapter code maps from the formats used by the external system only to what is required by the domain model. It has no other function.

This structure naturally follows the test pyramid guidelines. Fewer integration tests are required. Each integration test has only a small scope of behavior to test:

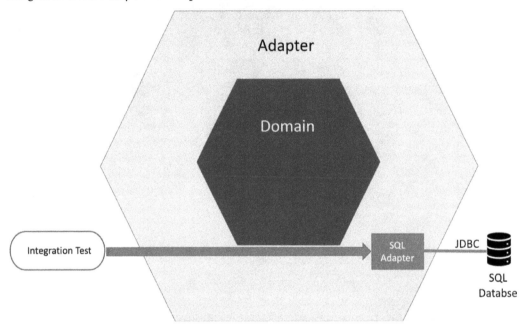

Figure 12.9 – Testing adapters

This style of testing verifies the adapter in isolation. It will require some end-to-end happy-path testing to show that the system as a whole has used the correct adapters.

User stories can be tested across the domain model

One benefit of having a domain model containing all the application logic is that we can test the logic of complete user stories. We can replace the adapters with test doubles to simulate typical responses from the external systems. We can then use FIRST unit tests to exercise complete user stories:

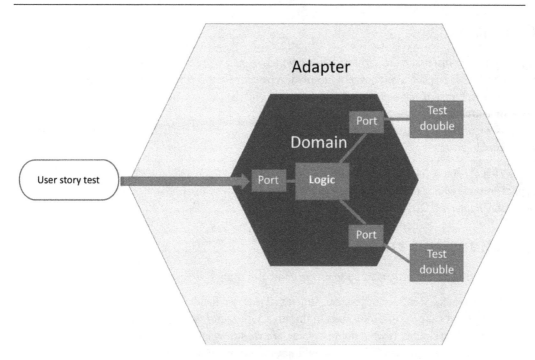

Figure 12.10 – Testing user stories

The advantages are the speed and repeatability of FIRST unit tests. In other approaches to structuring our code, we might only be able to exercise a user story as an end-to-end test in a test environment, with all associated disadvantages. Having the ability to test user story logic at the unit level – across the whole domain model – gives us a high degree of confidence that our application will satisfy the users' needs.

To ensure this confidence, we need the integration tests of the adapter layer, plus some end-to-end tests across selected user stories, confirming the application is wired and configured correctly as a whole. These higher-level tests do not need to be as detailed as the user story tests performed around the domain model.

Having a good set of user story tests around the domain model also enables large-scale refactoring within the domain model. We can have the confidence to restructure the inner hexagon guided by these broadly scoped user story tests.

This section has shown us how to relate the different kinds of tests in the test pyramid to the different layers in a hexagonal architecture.

Summary

This chapter has discussed the various stages at which we can write tests – before we write code, after we write code, or possibly even never. It has made a case for writing tests before code as providing the most value in terms of valid execution path coverage and developer ease. We went on to review how hexagonal architecture interacts with both TDD and the test pyramid, leading to an opportunity to bring user story testing into the realm of FIRST unit tests. This allows the fast and repeatable validation of the core logic driving our user stories.

In the next chapter – and throughout the third part of the book – we will return to building our Wordz application. We will be making full use of all the techniques we've learned so far. We will begin inside-out with *Chapter 13, Driving the Domain Layer*.

Questions and answers

1. Is writing tests shortly after code just as good as writing test-first TDD?

   Some research seems to suggest that, although it is very difficult to set up a controlled experiment with statistically significant results in this area. One factor we can consider concerns our own personal discipline. If we write tests later, are we sure we will cover everything necessary? I personally have concluded that I would not remember all I needed to cover and would need to make notes. Those notes are perhaps best captured in the form of test code, leading to a preference for test-first TDD.

2. How does hexagonal architecture affect TDD?

   Hexagonal architecture provides a clean separation between a pure, inner core of domain logic and the outside world. This allows us to mix and match the two schools of TDD knowing that there is a firm boundary in the design up to which we can code. The inner domain model supports entire use cases being unit-tested, as well as any fine-grained unit tests for detailed behavior we feel are necessary. External adapters naturally suit integration tests, but these tests don't have to cover much, as the logic relates to our domain lives in the inner domain model.

3. What happens if we abandon testing completely?

   We export the responsibility to the end user who will test it for us. We risk loss in revenue, reputation, and user retention. Sometimes, we cannot perfectly recreate the final environment in which the system will be used. In this case, making sure we have fully characterized and tested our system as closely as we can seems wise. We can at least minimize the known risks.

Further reading

- An explanation of the Cyclomatic Complexity metric: `https://en.wikipedia.org/wiki/Cyclomatic_complexity`

- *Continuous Delivery*, Jez Humble and Dave Farley, ISBN 978-0321601919

- *Working Effectively with Legacy Code*, Michael Feathers, ISBN 978-0131177055

- *Test-Driven Development by Example*, Kent Beck, ISBN 978-0321146533

- *Growing Object-Oriented Software, Guided by Tests*, Steve Freeman and Nat Pryce, ISBN 9780321503626

- `https://arxiv.org/pdf/1611.05994.pdf`

- *Why Research on Test-Driven Development is Inconclusive?*, Ghafari, Gucci, Gross, and Felderer: `https://arxiv.org/pdf/2007.09863.pdf`

Part 3: Real-World TDD

Part 3 is where we apply all the techniques we have learned to complete our application. Wordz is a web service that plays a word guessing game. We build on the core domain logic we have already built, adding storage via a Postgres database accessed using SQL and providing web access by implementing an HTTP REST API.

We will use integration tests to test-drive our database and API implementations, making use of test frameworks that simplify these tasks. In the final chapter of the book, we will bring everything together to confidently run our test-driven Wordz application.

This part has the following chapters:

- *Chapter 13, Driving the Domain Layer*
- *Chapter 14, Driving the Database Layer*
- *Chapter 15, Driving the Web Layer*

13

Driving the Domain Layer

We laid a lot of groundwork in previous chapters, covering a mixture of TDD techniques and software design approaches. Now we can apply those capabilities to build our Wordz game. We will be building on top of the useful code we have written throughout the book and working toward a well-engineered, well-tested design, written using the test-first approach.

Our goal for this chapter is to create the domain layer of our system. We will adopt the hexagonal architecture approach as described in *Chapter 9, Hexagonal Architecture – Decoupling External Systems*. The domain model will contain all our core application logic. This code will not be tied to details of any external system technologies such as SQL databases or web servers. We will create abstractions for these external systems and use test doubles to enable us to test-drive the application logic.

Using hexagonal architecture in this way allows us to write FIRST unit tests for complete user stories, which is something often requiring integration or end-to-end testing in other design approaches. We will write our domain model code by applying the ideas presented in the book so far.

In this chapter, we're going to cover the following main topics:

- Starting a new game
- Playing the game
- Ending the game

Technical requirements

The final code for this chapter can be found at `https://github.com/PacktPublishing/Test-Driven-Development-with-Java/tree/main/chapter13`.

Starting a new game

In this section, we will make a start by coding our game. Like every project, starting is usually quite difficult, with the first decision being simply where to begin. A reasonable approach is to find a user story that will begin to flesh out the structure of the code. Once we have a reasonable structure for an application, it becomes much easier to figure out where new code should be added.

Given this, we can make a good start by considering what needs to happen when we start a new game. This must set things up ready to play and so will force some critical decisions to be made.

The first user story to work on is starting a new game:

- As a player I want to start a new game so that I have a new word to guess

When we start a new game, we must do the following:

1. Select a word at random from the available words to guess

2. Store the selected word so that scores for guesses can be calculated

3. Record that the player may now make an initial guess

We will assume the use of hexagonal architecture as we code this story, meaning that any external system will be represented by a port in the domain model. With this in mind, we can create our first test and take it from there.

Test-driving starting a new game

In terms of a general direction, using hexagonal architecture means we are free to use an outside-in approach with TDD. Whatever design we come up with for our domain model, none of it is going to involve difficult-to-test external systems. Our unit tests are assured to be **FIRST – fast, isolated, repeatable, self-checking, and timely**.

Importantly, we can write unit tests that cover the entire logic needed for a user story. If we wrote code that is bound to external systems – for example, it contained SQL statements and connected to a database – we would need an integration test to cover a user story. Our choice of hexagonal architecture frees us from that.

On a tactical note, we will reuse classes that we have already test-driven, such as `class WordSelection`, `class Word`, and `class Score`. We will reuse existing code and third-party libraries whenever an opportunity presents itself.

Our starting point is to write a test to capture our design decisions related to starting a new game:

1. We will start with a test called `NewGameTest`. This test will act across the domain model to drive out our handling of everything we need to do to start a new game:

```
package com.wordz.domain;


public class NewGameTest {
}
```

2. For this test, we will start with the Act step first. We are assuming hexagonal architecture, so the design goal of the Act step is to design the port that handles the request to start a new game. In hexagonal architecture, a port is the piece of code that allows some external system to connect with the domain model. We begin by creating a class for our port:

```
package com.wordz.domain;


public class NewGameTest {
    void startsNewGame() {
        var game = new Game();
    }
}
```

The key design decision here is to create a `controller` class to handle the request to start a game. It is a controller in the sense of the original Gang of Four's *Design Patterns* book – a domain model object that will orchestrate other domain model objects. We will let the IntelliJ IDE create the empty Game class:

```
package com.wordz.domain;


public class Game {
}
```

That's another advantage of TDD. When we write the test first, we give our IDE enough information to be able to generate boilerplate code for us. We enable the IDE autocomplete feature to really help us. If your IDE cannot autogenerate code after having written the test, consider upgrading your IDE.

3. The next step is to add a `start()` method on the controller class to start a new game. We need to know which player we are starting a game for, so we pass in a `Player` object. We write the Act step of our test:

```
public class NewGameTest {
    @Test
    void startsNewGame() {
        var game = new Game();
        var player = new Player();


        game.start(player);
    }
}
```

We allow the IDE to generate the method in the controller:

```
public class Game {
    public void start(Player player) {
    }
}
```

Tracking the progress of the game

The next design decisions concern the expected outcome of starting a new game for a player. There are two things that need to be recorded:

- The selected word that the player attempts to guess
- That we expect their first guess next

The selected word and current attempt number will need to persist somewhere. We will use the repository pattern to abstract that. Our repository will need to manage some domain objects. Those objects will have the single responsibility of tracking our progress in a game.

Already, we see a benefit of TDD in terms of rapid design feedback. We haven't written too much code yet, but already, it seems like the new class needed to track game progress would best be called class Game. However, we already have a class Game, responsible for starting a new game. TDD is providing feedback on our design – that our names and responsibilities are mismatched.

We must choose one of the following options to proceed:

- Keep our existing class Game as it is. Call this new class something such as Progress or Attempt.
- Change the start() method to a static method – a method that applies to all instances of a class.
- Rename class Game to something that better describes its responsibility. Then, we can create a new class Game to hold current player progress.

The static method option is unappealing. When using object-oriented programming in Java, static methods rarely seem as good a fit as simply creating another object that manages all the relevant instances. The static method becomes a normal method on this new object. Using class Game to represent progress through a game seems to result in more descriptive code. Let's go with that approach.

1. Use the IntelliJ IDEA IDE to refactor/rename class Game class Wordz, which represents the entry point into our domain model. We also rename the local variable game to match:

```
public class NewGameTest {
    @Test
```

```
        void startsNewGame() {
            var wordz = new Wordz();
            var player = new Player();

            wordz.start(player);
        }
    }
```

The name of the NewGameTest test is still good. It represents the user story we are testing and is not related to any class names. The production code has been refactored by the IDE as well:

```
public class Wordz {

    public void start(Player player) {
    }
}
```

2. Use the IDE to refactor/rename the start() method newGame(). This seems to better describe the responsibility of the method, in the context of a class named Wordz:

```
public class NewGameTest {
    @Test
    void startsNewGame() {
        var wordz = new Wordz();
        var player = new Player();

        wordz.newGame(player);
    }
}
```

The class Wordz production code also has the method renamed.

3. When we start a new game, we need to select a word to guess and start the sequence of attempts the player has. These facts need to be stored in a repository. Let's create the repository first. We will call it interface GameRepository and add Mockito @Mock support for it in our test:

```
package com.wordz.domain;

import org.junit.jupiter.api.Test;
import org.junit.jupiter.api.extension.ExtendWith;
import org.mockito.Mock;
```

```
import org.mockito.junit.jupiter.MockitoExtension;

@ExtendWith(MockitoExtension.class)
public class NewGameTest {
    @Mock
    private GameRepository gameRepository;

    @InjectMocks
    private Wordz wordz;

    @Test
    void startsNewGame() {
        var player = new Player();

        wordz.newGame(player);
    }
}
```

We add the @ExtendWith annotation to the class to enable the Mockito library to automatically create test doubles for us. We add a gameRepository field, which we annotated as a Mockito @Mock. We use the @InjectMocks convenience annotation built into Mockito to automatically inject this dependency into the Wordz constructor.

4. We allow the IDE to create an empty interface for us:

```
package com.wordz.domain;

public interface GameRepository {
}
```

5. For the next step, we will confirm that gameRepository gets used. We decide to add a create() method on the interface, which takes a class Game object instance as its only parameter. We want to inspect that object instance of class Game, so we add an argument captor. This allows us to assert on the game data contained in that object:

```
public class NewGameTest {
    @Mock
    private GameRepository gameRepository;
    @Test
    void startsNewGame() {
```

```
        var player = new Player();

        wordz.newGame(player);

        var gameArgument =
              ArgumentCaptor.forClass(Game.class)
        verify(gameRepository)
          .create(gameArgument.capture());
        var game = gameArgument.getValue();

        assertThat(game.getWord()).isEqualTo("ARISE");
        assertThat(game.getAttemptNumber()).isZero();
        assertThat(game.getPlayer()).isSameAs(player);
      }
   }
```

A good question is why we are asserting against those particular values. The reason is that we are going to cheat when we add the production code and *fake it until we make it*. We will return a Game object that hardcodes these values as a first step. We can then work in small steps. Once the cheat version makes the test pass, we can refine the test and test-drive the code to fetch the word for real. Smaller steps provide more rapid feedback. Rapid feedback enables better decision-making.

Note on using getters in the domain model

The Game class has getXxx() methods, known as *getters* in Java terminology, for every one of its private fields. These methods break the encapsulation of data.

This is generally not recommended. It can lead to important logic being placed into other classes – a code smell known as a foreign method. Object-oriented programming is all about co-locating logic and data, encapsulating both. Getters should be few and far between. That does not mean we should never use them, however.

In this case, the single responsibility of class Game is to transfer the current state of the game being played to GameRepository. The most direct way of implementing this is to add getters to the class. Writing simple, clear code beats following rules dogmatically.

Another reasonable approach is to add a getXxx() **diagnostic method** at package-level visibility purely for testing. Check with the team that this is not part of the public API and do not use it in production code. It is more important to get the code correct than obsess over design trivia.

6. We create empty methods for these new getters using the IDE. The next step is to run NewGameTest and confirm that it fails:

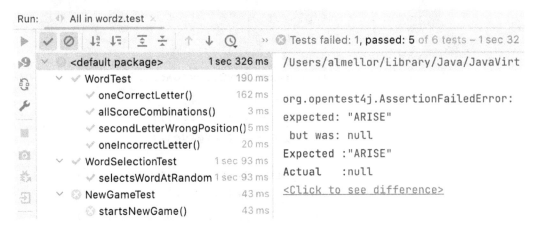

Figure 13.1 – Our failing test

7. This is enough for us to write some more production code:

```
package com.wordz.domain;

public class Wordz {
    private final GameRepository gameRepository;

    public Wordz(GameRepository gr) {
        this.gameRepository = gr;
    }

    public void newGame(Player player) {
        var game = new Game(player, "ARISE", 0);
        gameRepository.create(game);
    }
}
```

We can rerun NewGameTest and watch it pass:

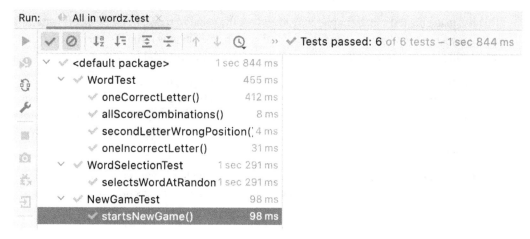

Figure 13.2 – The test passes

The test now passes. We can move from our red-green phase to thinking about refactoring. The thing that jumps out immediately is just how unreadable that ArgumentCaptor code is in the test. It contains too much detail about the mechanics of mocking and not enough detail about why we are using that technique. We can clarify that by extracting a well-named method.

8. Extract the getGameInRepository() method for clarity:

```
@Test
void startsNewGame() {
    var player = new Player();

    wordz.newGame(player);

    Game game = getGameInRepository();
    assertThat(game.getWord()).isEqualTo("ARISE");
    assertThat(game.getAttemptNumber()).isZero();
    assertThat(game.getPlayer()).isSameAs(player);
}

private Game getGameInRepository() {
    var gameArgument
        = ArgumentCaptor.forClass(Game.class)
    verify(gameRepository)
            .create(gameArgument.capture());
    return gameArgument.getValue();
}
```

That has made the test much simpler to read and see the usual Arrange, Act, and Assert pattern in it. It is a simple test by nature and should read as such. We can now rerun the test and confirm that it still passes. It does, and we are satisfied that our refactoring did not break anything.

That completes our first test – a job well done! We're making good progress here. It always feels good to me to see a test go green, and that feeling never gets old. This test is essentially an end-to-end test of a user story, acting only on the domain model. Using hexagonal architecture enables us to write tests that cover the details of our application logic, while avoiding the need for test environments. We get faster-running, more stable tests as a result.

There is more work to do in our next test, as we need to remove the hardcoded creation of the Game object. In the next section, we will address this by triangulating the word selection logic. We design the next test to drive out the correct behavior of selecting a word at random.

Triangulating word selection

The next task is to remove the cheating that we used to make the previous test pass. We hardcoded some data when we created a Game object. We need to replace that with the correct code. This code must select a word at random from our repository of known five-letter words.

1. Add a new test to drive out the behavior of selecting a random word:

    ```
    @Test
    void selectsRandomWord() {


    }
    ```

2. Random word selection depends on two external systems – the database that holds the words to choose from and a source of random numbers. As we are using hexagonal architecture, the domain layer cannot access those directly. We will represent them with two interfaces – the ports to those systems. For this test, we will use **Mockito** to create stubs for those interfaces:

    ```
    @ExtendWith(MockitoExtension.class)
    public class NewGameTest {
        @Mock
        private GameRepository gameRepository;

        @Mock
        private WordRepository wordRepository ;

        @Mock
        private RandomNumbers randomNumbers ;
    ```

```
@InjectMocks
private Wordz wordz;
```

This test introduces two new collaborating objects to class Wordz. These are instances of any valid implementations of both interface WordRepository and interface RandomNumbers. We need to inject those objects into the Wordz object to make use of them.

3. Using dependency injection, inject the two new interface objects into the class Wordz constructor:

```
public class Wordz {
    private final GameRepository gameRepository;
    private final WordSelection wordSelection ;

    public Wordz(GameRepository gr,
                 WordRepository wr,
                 RandomNumbers rn) {
        this.gameRepository = gr;
        this.wordSelection = new WordSelection(wr, rn);
    }
```

We've added two parameters to the constructor. We do not need to store them directly as fields. Instead, we use the previously created class WordSelection. We create a WordSelection object and store it in a field called wordSelection. Note that our earlier use of @InjectMocks means that our test code will automatically pass in the mock objects to this constructor, without further code changes. It is very convenient.

4. We set up the mocks. We want them to simulate the behavior we expect from interface WordRepository when we call the fetchWordByNumber() method and interface RandomNumbers when we call next():

```
@Test
void selectsRandomWord() {
    when(randomNumbers.next(anyInt())).thenReturn(2);
    when(wordRepository.fetchWordByNumber(2))
            .thenReturn("ABCDE");
}
```

This will set up our mocks so that when next() is called, it will return the word number 2 every time, as a test double for the random number that will be produced in the full application. When fetchWordByNumber() is then called with 2 as an argument, it will return the word with word number 2, which will be "ABCDE" in our test. Looking at that code, we can add clarity by using a local variable instead of that magic number 2. To future readers of the code, the link between random number generator output and word repository will be more obvious:

```
@Test
void selectsRandomWord() {
    int wordNumber = 2;

    when(randomNumbers.next(anyInt()))
        .thenReturn(wordNumber);

    when(wordRepository
        .fetchWordByNumber(wordNumber))
            .thenReturn("ABCDE");
}
```

5. That still looks too detailed once again. There is too much emphasis on mocking mechanics and too little on what the mocking represents. Let's extract a method to explain why we are setting up this stub. We will also pass in the word we want to be selected. That will help us more easily understand the purpose of the test code:

```
@Test
void selectsRandomWord() {
    givenWordToSelect("ABCDE");
}

private void givenWordToSelect(String wordToSelect){
    int wordNumber = 2;

    when(randomNumbers.next(anyInt()))
            .thenReturn(wordNumber);

    when(wordRepository
            .fetchWordByNumber(wordNumber))
            .thenReturn(wordToSelect);

}
```

6. Now, we can write the assertion to confirm that this word is passed down to the `gameRepository` `create()` method – we can reuse our `getGameInRepository()` assert helper method:

```
@Test
void selectsRandomWord() {
    givenWordToSelect("ABCDE");

    var player = new Player();
    wordz.newGame(player);

    Game game = getGameInRepository();
    assertThat(game.getWord()).isEqualTo("ABCDE");
}
```

This follows the same approach as the previous test, `startsNewGame`.

7. Watch the test fail. Write production code to make the test pass:

```
public void newGame(Player player) {
    var word = wordSelection.chooseRandomWord();
    Game game = new Game(player, word, 0);
    gameRepository.create(game);
}
```

8. Watch the new test pass and then run all tests:

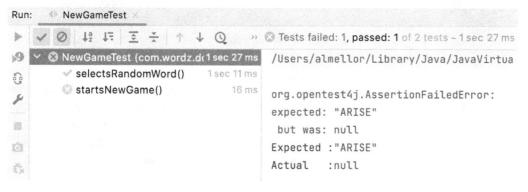

Figure 13.3 – Original test failing

Our initial test has now failed. We've broken something during our latest code change. TDD has kept us safe by providing a regression test for us. What has happened is that after removing the hardcoded word `"ARISE"` that the original test relied on, it fails. The correct solution is to add the required mock setup to our original test. We can reuse our `givenWordToSelect()` helper method to do this.

9. Add the mock setup to the original test:

```java
@Test
void startsNewGame() {
    var player = new Player();
    givenWordToSelect("ARISE");

    wordz.newGame(player);

    Game game = getGameInRepository();
    assertThat(game.getWord()).isEqualTo("ARISE");
    assertThat(game.getAttemptNumber()).isZero();
    assertThat(game.getPlayer()).isSameAs(player);
}
```

10. Rerun all tests and confirm that they all pass:

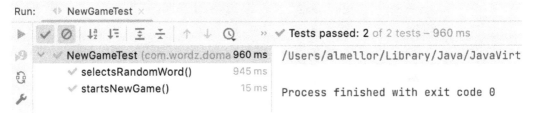

Figure 13.4 – All tests passing

We've test-driven our first piece of code to start a new game, with a randomly selected word to guess, and made the tests pass. Before we move on, it is time to consider what – if anything – we should refactor. We have been tidying the code as we write it, but there is one glaring feature. Take a look at the two tests. They seem very similar now. The original test has become a superset of the one we used to test-drive adding the word selection. The `selectsRandomWord()` test is a **scaffolding test** that no longer serves a purpose. There's only one thing to do with code like that – remove it. As a minor readability improvement, we can also extract a constant for the `Player` variable:

1. Extract a constant for the `Player` variable:

```java
private static final Player PLAYER = new Player();

@Test
```

```
void startsNewGame() {
    givenWordToSelect("ARISE");

    wordz.newGame(PLAYER);

    Game game = getGameInRepository();
    assertThat(game.getWord()).isEqualTo("ARISE");
    assertThat(game.getAttemptNumber()).isZero();
    assertThat(game.getPlayer()).isSameAs(PLAYER);
}
```

2. We'll run all the tests after this to make sure that they all still pass and that
    `selectsRandomWord()` has gone.

Figure 13.5 – All tests passing

That's it! We have test-driven out all the behavior we need to start a game. It's a significant achievement because that test covers a complete user story. All the domain logic has been tested and is known to be working. The design looks straightforward. The test code is a clear specification of what we expect our code to do. This is great progress.

Following this refactoring, we can move on to the next development task – code that supports playing the game.

Playing the game

In this section, we will build the logic to play the game. The gameplay consists of making a number of guesses at the selected word, reviewing the score for that guess, and having another guess. The game ends either when the word has been guessed correctly or when the maximum number of allowed attempts has been made.

We'll begin by assuming that we are at the start of a typical game, about to make our first guess. We will also assume that this guess is not completely correct. This allows us to defer decisions about end-of-the-game behavior, which is a good thing, as we have enough to decide already.

Designing the scoring interface

The first design decision we must take is what we need to return following a guess at the word. We need to return the following information to the user:

- The score for the current guess
- Whether or not the game is still in play or has ended
- Possibly the previous history of scoring for each guess
- Possibly a report of user input errors

Clearly, the most important information for the player is the score for the current guess. Without that, the game cannot be played. As the game has a variable length – ending when either the word has been guessed, or when a maximum number of guesses has been attempted – we need an indicator that another guess will be allowed.

The idea behind returning the history of scores for previous guesses is that it might help the consumer of our domain model – ultimately, a user interface of some sort. If we return only the score for the current guess, the user interface will most likely need to retain its own history of scores, in order to present them properly. If we return the entire history of scores for this game, that information is easily available. A good rule of thumb in software is to follow the **you ain't gonna need it** (**YAGNI**) principle. As there is no requirement for a history of scores, we won't build that at this stage.

The last decision we need to write our test is to think about the programming interface we want for this. We will choose an `assess()` method on `class Wordz`. It will accept `String`, which is the current guess from the player. It will return `record`, which is a modern Java (since Java 14) way of indicating a pure data structure is to be returned:

We've now got enough to write a test. We'll make a new test for all guess-related behavior called `class GuessTest`. The test looks like this:

```
@ExtendWith(MockitoExtension.class)
public class GuessTest {
    private static final Player PLAYER = new Player();
    private static final String CORRECT_WORD = "ARISE";
    private static final String WRONG_WORD = "RXXXX";

    @Mock
    private GameRepository gameRepository;

    @InjectMocks
    private Wordz wordz;
```

```
    @Test
    void returnsScoreForGuess() {
        givenGameInRepository(
                    Game.create(PLAYER, CORRECT_WORD));

        GuessResult result = wordz.assess(PLAYER, WRONG_WORD);

        Letter firstLetter = result.score().letter(0);
        assertThat(firstLetter)
            .isEqualTo(Letter.PART_CORRECT);
    }

    private void givenGameInRepository(Game game) {
        when(gameRepository
            .fetchForPlayer(eq(PLAYER)))
                .thenReturn(Optional.of(game));
    }
}
```

There are no new TDD techniques in the test. It drives out the calling interface for our new assess()
method. We've used the static constructor idiom to create the game object using Game.create().
This method has been added to class Game:

```
static Game create(Player player, String correctWord) {
    return new Game(player, correctWord, 0, false);
}
```

This clarifies the information necessary to create a new game. To get the test to compile, we create
record GuessResult:

```
package com.wordz.domain;

import java.util.List;

public record GuessResult(
        Score score,
        boolean isGameOver
) { }
```

We can make the test pass by writing the production code for the assess() method in class Wordz. To do that, we will reuse the class Word class that we have already written:

```java
public GuessResult assess(Player player, String guess) {
    var game = gameRepository.fetchForPlayer(player);
    var target = new Word(game.getWord());
    var score = target.guess(guess);
    return new GuessResult(score, false);
}
```

The assertion checks only that the score for the first letter is correct. This is intentionally a weak test. The detailed testing for scoring behavior is done in class WordTest, which we wrote previously. The test is described as weak, as it does not fully test the returned score, only the first letter of it. Strong testing of the scoring logic happens elsewhere, in class WordTest. The weak test here confirms we have something capable of scoring at least one letter correctly and is enough for us to test-drive the production code. We avoid duplicating tests here.

Running the test shows that it passes. We can review the test code and production code to see whether refactoring will improve their design. At this point, nothing needs our urgent attention. We can move on to tracking progress through the game.

Triangulating game progress tracking

We need to track the number of guesses that have been made so that we can end the game after a maximum number of attempts. Our design choice is to update the attemptNumber field in the Game object and then store it in GameRepository:

1. We add a test to drive this code out:

```java
@Test
void updatesAttemptNumber() {
    givenGameInRepository(
            Game.create(PLAYER, CORRECT_WORD));

    wordz.assess(PLAYER, WRONG_WORD);

    var game = getUpdatedGameInRepository();
    assertThat(game.getAttemptNumber()).isEqualTo(1);
}

private Game getUpdatedGameInRepository() {
```

```
ArgumentCaptor<Game> argument
        = ArgumentCaptor.forClass(Game.class);
verify(gameRepository).update(argument.capture());
return argument.getValue();
}
```

This test introduces a new method, update(), into our interface GameRepository, responsible for writing the latest game information to storage. The Assert step uses a Mockito ArgumentCaptor to inspect the Game object that we pass into update(). We have written a getUpdatedGameInRepository() method to deemphasize the inner workings of how we check what was passed to the gameRepository.update() method. assertThat() in the test verifies that attemptNumber has been incremented. It started at zero, due to us creating a new game, and so the expected new value is 1. This is the desired behavior for tracking an attempt to guess the word:

2. We add the update() method to the GameRepository interface:

```
package com.wordz.domain;

public interface GameRepository {
    void create(Game game);
    Game fetchForPlayer(Player player);
    void update(Game game);
}
```

3. We add the production code to the assess() method in class Wordz to increment attemptNumber and call update():

```
public GuessResult assess(Player player, String guess) {
    var game = gameRepository.fetchForPlayer(player);
    game.incrementAttemptNumber();
    gameRepository.update(game);
    var target = new Word(game.getWord());
    var score = target.guess(guess);
    return new GuessResult(score, false);
}
```

4. We add the incrementAttemptNumber() method to class Game:

```
public void incrementAttemptNumber() {
    attemptNumber++;
}
```

The test now passes. We can think about any refactoring improvements we want to make. There are two things that seem to stand out:

- The duplicated test setup between class NewGameTest and class GuessTest.

  At this stage, we can live with this duplication. The options are to combine both tests into the same test class, to extend a common test base class, or to use composition. None of them seem likely to aid readability much. It seems quite nice to have the two different test cases separate for now.

- The three lines inside the assess() method must always be called as a unit when we attempt another guess. It is possible to forget to call one of these, so it seems better to refactor to eliminate that possible error. We can refactor like this:

```
public GuessResult assess(Player player, String guess) {
    var game = gameRepository.fetchForPlayer(player);
    Score score = game.attempt( guess );
    gameRepository.update(game);
    return new GuessResult(score, false);
}
```

We move the code that used to be here into the newly created method: attempt() on class Game:

```
public Score attempt(String latestGuess) {
    attemptNumber++;
    var target = new Word(targetWord);
    return target.guess(latestGuess);
}
```

Renaming the method argument from guess to latestGuess improves readability.

That completes the code needed to take a guess at the word. Let's move on to test-driving the code we will need to detect when a game has ended.

Ending the game

In this section, we will complete the tests and production code we need to drive out detecting the end of a game. This will happen when we do either of the following:

- Guess the word correctly
- Make our final allowed attempt, based on a maximum number

We can make a start by coding the end-of-game detection when we guess the word correctly.

Responding to a correct guess

In this case, the player guesses the target word correctly. The game is over, and the player is awarded a number of points, based on how few attempts were needed before the correct guess was made. We need to communicate that the game is over and how many points have been awarded, leading to two new fields in our class GuessResult. We can add a test to our existing class GuessTest as follows:

```
@Test
void reportsGameOverOnCorrectGuess() {
    var player = new Player();
    Game game = new Game(player, "ARISE", 0);
    when(gameRepository.fetchForPlayer(player))
                        .thenReturn(game);
    var wordz = new Wordz(gameRepository,
                          wordRepository, randomNumbers);

    var guess = "ARISE";
    GuessResult result = wordz.assess(player, guess);

    assertThat(result.isGameOver()).isTrue();
}
```

This drives out both a new isGameOver()accessor in class GuessResult and the behavior to make that true:

```
public GuessResult assess(Player player, String guess) {
    var game = gameRepository.fetchForPlayer(player);
    Score score = game.attempt( guess );
    if (score.allCorrect()) {
        return new GuessResult(score, true);
    }

    gameRepository.update(game);
    return new GuessResult(score, false);
}
```

This itself drives out two new tests in class WordTest:

```
@Test
void reportsAllCorrect() {
```

```
        var word = new Word("ARISE");
        var score = word.guess("ARISE");
        assertThat(score.allCorrect()).isTrue();
    }


    @Test
    void reportsNotAllCorrect() {
        var word = new Word("ARISE");
        var score = word.guess("ARI*E");
        assertThat(score.allCorrect()).isFalse();
    }
```

These themselves drive out an implementation in class Score:

```
    public boolean allCorrect() {
        var totalCorrect = results.stream()
                .filter(letter -> letter == Letter.CORRECT)
                .count();


        return totalCorrect == results.size();
    }
```

With this, we have a valid implementation for the isGameOver accessor in record GuessResult. All tests pass. Nothing seems to need refactoring. We'll move on to the next test.

Triangulating the game over due to too many incorrect guesses

The next test will drive out the response to exceeding the maximum number of guesses allowed in a game:

```
    @Test
    void gameOverOnTooManyIncorrectGuesses() {
        int maximumGuesses = 5;
        givenGameInRepository(
                Game.create(PLAYER, CORRECT_WORD,
                    maximumGuesses-1));


        GuessResult result = wordz.assess(PLAYER, WRONG_WORD);
```

```
    assertThat(result.isGameOver()).isTrue();
}
```

This test sets up gameRepository to allow one, final guess. It then sets up the guess to be incorrect. We assert that isGameOver() is true in this case. The test fails initially, as desired. We add an extra static constructor method in class Game to specify an initial number of attempts.

We add the production code to end the game based on a maximum number of guesses:

```
public GuessResult assess(Player player, String guess) {
    var game = gameRepository.fetchForPlayer(player);
    Score score = game.attempt( guess );
    if (score.allCorrect()) {
        return new GuessResult(score, true);
    }


    gameRepository.update(game);
    return new GuessResult(score,
                           game.hasNoRemainingGuesses());
}
```

We add this decision support method to class Game:

```
public boolean hasNoRemainingGuesses() {
    return attemptNumber == MAXIMUM_NUMBER_ALLOWED_GUESSES;
}
```

All our tests now pass. There is something suspicious about the code, however. It has been very finely tuned to work only if a guess is correct and within the allowed number of guesses, or when the guess is incorrect and exactly at the allowed number. It's time to add some boundary condition tests and double-check our logic.

Triangulating response to guess after game over

We need a couple more tests around the boundary conditions of the game over detection. The first one drives out the response to an incorrect guess being submitted after a correct guess:

```
@Test
void rejectsGuessAfterGameOver(){
    var gameOver = new Game(PLAYER, CORRECT_WORD,
                    1, true);
```

```
            givenGameInRepository( gameOver );

            GuessResult result = wordz.assess(PLAYER, WRONG_WORD);

            assertThat(result.isError()).isTrue();
    }
```

There are a couple of design decisions captured in this test:

- Once the game ends, we record this in a new field, isGameOver, in class Game.
- This new field will need to be set whenever the game ends. We will need more tests to drive that behavior out.
- We will use a simple error-reporting mechanism – a new field, isError, in class GuessResult.

This leads to a bit of automated refactoring to add the fourth parameter to the class Game constructor. Then, we can add code to make the test pass:

```
public GuessResult assess(Player player, String guess) {
    var game = gameRepository.fetchForPlayer(player);

    if(game.isGameOver()) {
        return GuessResult.ERROR;
    }

    Score score = game.attempt( guess );
    if (score.allCorrect()) {
        return new GuessResult(score, true, false);
    }

    gameRepository.update(game);
    return new GuessResult(score,
                    game.hasNoRemainingGuesses(), false);
}
```

The design decision here is that as soon as we fetch the Game object, we check whether the game was previously marked as being over. If so, we report an error and we're done. It's simple and crude but adequate for our purposes. We also add a static constant, GuessResult.ERROR, for readability:

```
public static final GuessResult ERROR
                 = new GuessResult(null, true, true);
```

One consequence of this design decision is that we must update GameRepository whenever the Game.isGameOver field changes to true. An example of one of these tests is this:

```
@Test
void recordsGameOverOnCorrectGuess(){
    givenGameInRepository(Game.create(PLAYER, CORRECT_WORD));

    wordz.assess(PLAYER, CORRECT_WORD);

    Game game = getUpdatedGameInRepository();
    assertThat(game.isGameOver()).isTrue();
}
```

Here is the production code to add that recording logic:

```
public GuessResult assess(Player player, String guess) {
    var game = gameRepository.fetchForPlayer(player);

    if (game.isGameOver()) {
        return GuessResult.ERROR;
    }

    Score score = game.attempt( guess );
    if (score.allCorrect()) {
        game.end();
        gameRepository.update(game);

        return new GuessResult(score, true, false);
    }

    gameRepository.update(game);
    return new GuessResult(score,
```

```
                          game.hasNoRemainingGuesses(), false);
    }
```

We need another test to drive out the recording of game over when we run out of guesses. That will lead to a change in the production code. Those changes can be found in GitHub at the link given at the start of this chapter. They are very similar to the ones made previously.

Finally, let's review our design and see whether we can improve it still further.

Reviewing our design

We've been making small, tactical refactoring steps as we write the code, which is always a good idea. Like gardening, it is far easier to keep the garden tidy if we pull up weeds before they grow. Even so, it is worth taking a holistic look at the design of our code and tests before we move on. We may never get the chance to touch this code again, and it has our name on it. Let's make it something that we are proud of and that will be safe and simple for our colleagues to work with in the future.

The tests we've already written enable us great latitude in refactoring. They have avoided testing specific implementations, instead testing desired outcomes. They also test larger units of code – in this case, the domain model of our hexagonal architecture. As a result, without changing any tests, it is possible to refactor our class Wordz to look like this:

```
package com.wordz.domain;

public class Wordz {
    private final GameRepository gameRepository;
    private final WordSelection selection ;

    public Wordz(GameRepository repository,
                 WordRepository wordRepository,
                 RandomNumbers randomNumbers) {
        this.gameRepository = repository;
        this.selection =
            new WordSelection(wordRepository, randomNumbers);
    }

    public void newGame(Player player) {
        var word = wordSelection.chooseRandomWord();
        gameRepository.create(Game.create(player, word));
    }
```

Our refactored `assess()` method now looks like this:

```
public GuessResult assess(Player player, String guess) {
    Game game = gameRepository.fetchForPlayer(player);

    if(game.isGameOver()) {
        return GuessResult.ERROR;
    }

    Score score = game.attempt( guess );

    gameRepository.update(game);
    return new GuessResult(score,
                              game.isGameOver(), false);
    }
}
```

That's looking simpler. The `class GuessResult` constructor code now stands out as being particularly ugly. It features the classic anti-pattern of using multiple Boolean flag values. We need to clarify what the different combinations actually mean, to simplify creating the object. One useful approach is to apply the static constructor idiom once more:

```
package com.wordz.domain;

public record GuessResult(
        Score score,
        boolean isGameOver,
        boolean isError
) {
    static final GuessResult ERROR
        = new GuessResult(null, true, true);

    static GuessResult create(Score score,
                                boolean isGameOver) {
        return new GuessResult(score, isGameOver, false);
    }
}
```

This simplifies the assess() method by eliminating the need to understand that final Boolean flag:

```
public GuessResult assess(Player player, String guess) {
    Game game = gameRepository.fetchForPlayer(player);

    if(game.isGameOver()) {
        return GuessResult.ERROR;
    }

    Score score = game.attempt( guess );

    gameRepository.update(game);

    return GuessResult.create(score, game.isGameOver());
}
```

Another improvement to aid understanding concerns creating new instances of class Game. The rejectsGuessAfterGameOver() test uses Boolean flag values in a four-argument constructor to set the test up in a game-over state. Let's make the goal of creating a game-over state explicit. We can make the Game constructor private, and increase the visibility of the end() method, which is already used to end a game. Our revised test looks like this:

```
@Test
void rejectsGuessAfterGameOver(){
    var game = Game.create(PLAYER, CORRECT_WORD);
    game.end();
    givenGameInRepository( game );

    GuessResult result = wordz.assess(PLAYER, WRONG_WORD);

    assertThat(result.isError()).isTrue();
}
```

The Arrange step is now more descriptive. The four-argument constructor is no longer accessible, steering future development to use the safer, more descriptive static constructor methods. This improved design helps prevent defects from being introduced in the future.

We have made great progress in this chapter. Following these final refactoring improvements, we have an easily readable description of the core logic of our game. It is fully backed by FIRST unit tests. We have even achieved a meaningful 100% code coverage of lines of code executed by our tests. This is shown in the IntelliJ code coverage tool:

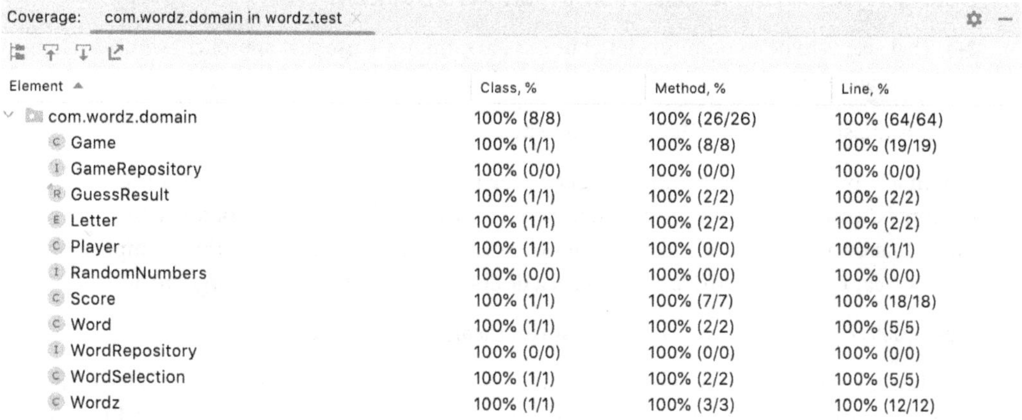

Element ▲	Class, %	Method, %	Line, %
∨ ▣ com.wordz.domain	100% (8/8)	100% (26/26)	100% (64/64)
Ⓒ Game	100% (1/1)	100% (8/8)	100% (19/19)
Ⓘ GameRepository	100% (0/0)	100% (0/0)	100% (0/0)
Ⓡ GuessResult	100% (1/1)	100% (2/2)	100% (2/2)
Ⓔ Letter	100% (1/1)	100% (2/2)	100% (2/2)
Ⓒ Player	100% (1/1)	100% (0/0)	100% (1/1)
Ⓘ RandomNumbers	100% (0/0)	100% (0/0)	100% (0/0)
Ⓒ Score	100% (1/1)	100% (7/7)	100% (18/18)
Ⓒ Word	100% (1/1)	100% (2/2)	100% (5/5)
Ⓘ WordRepository	100% (0/0)	100% (0/0)	100% (0/0)
Ⓒ WordSelection	100% (1/1)	100% (2/2)	100% (5/5)
Ⓒ Wordz	100% (1/1)	100% (3/3)	100% (12/12)

Coverage: com.wordz.domain in wordz.test ×

Figure 13.6 – Code coverage report

That's the core of our game finished. We can start a new game, play a game, and end a game. The game can be developed further to include features such as awarding a points score based on how quickly the word was guessed and a high score table for players. These would be added using the same techniques we have been applying throughout this chapter.

Summary

We've covered a lot of ground in this chapter. We have used TDD to drive out the core application logic for our Wordz game. We have taken small steps and used triangulation to steadily drive more details into our code implementation. We have used hexagonal architecture to enable us to use FIRST unit tests, freeing us from cumbersome integration tests with their test environments. We have employed test doubles to replace difficult-to-control objects, such as the database and random number generation.

We built up a valuable suite of unit tests that are decoupled from specific implementations. This enabled us to refactor the code freely, ending up with a very nice software design, based on the SOLID principles, which will reduce maintenance efforts significantly.

We finished with a meaningful code coverage report that showed 100% of the lines of production code were executed by our tests, giving us a high degree of confidence in our work.

Next, in *Chapter 14, Driving the Database Layer*, we will write the database adapter along with an integration test to implement our `GameRepository`, using the Postgres database.

Questions and answers

1. Does every method in every class have to have its own unit test?

   No. That seems to be a common view, but it is harmful. If we use that approach, we are locking in the implementation details and will not be able to refactor without breaking tests.

2. What is the significance of 100% code coverage when running our tests?

   Not much, by itself. It simply means that all the lines of code in the units under the test were executed during the test run. For us, it means a little more due to our use of test-first TDD. We know that every line of code was driven by a meaningful test of behavior that is important to our application. Having 100% coverage is a double-check that we didn't forget to add a test.

3. Does 100% code coverage during the test run mean we have perfect code?

   No. Testing can only reveal the presence of defects, never their absence. We can have 100% coverage with very low-quality code in terms of readability and edge case handling. It is important to not attach too much importance to code coverage metrics. For TDD, they serve as a cross-check that we haven't missed any boundary condition tests.

4. Is all this refactoring normal?

   Yes. TDD is all about rapid feedback loops. Feedback helps us explore design ideas and change our minds as we uncover better designs. It frees us from the tyranny of having to understand every detail – somehow – before we start work. We discover a design by doing the work and have working software to show for it at the end.

Further reading

- AssertJ documentation – read more about the various kinds of assertion matchers built into AssertJ, as well as details on how to create custom assertions here: `https://assertj.github.io/doc/`.

- *Refactoring – Improving the Design of Existing Code*, Martin Fowler (first edition), ISBN 9780201485677:

   The bulk of our work in TDD is refactoring code, continuously providing a good-enough design to support our new features. This book contains excellent advice on how to approach refactoring in a disciplined, step-by-step way.

   The first edition of the book uses Java for all its examples, so is more useful to us than the JavaScript-based second edition.

- *Design Patterns – Elements of Reusable Object-Oriented Software*, Gamma, Helm, Vlissides, Johnson, ISBN 9780201633610:

  A landmark book that cataloged common combinations of classes that occur in object-oriented software. Earlier in the chapter, we used a controller class. This is described as a façade pattern, in the terms of this book. The listed patterns are free of any kind of framework or software layer and so are very useful in building the domain model of hexagonal architecture.

14

Driving the Database Layer

In this chapter, we will implement a database adapter for one of our ports in the domain model, represented by the WordRepository interface. This will allow our domain model to fetch words to guess from a real database, in this case, using the popular open source database **Postgres**. We will test-drive both the database setup and the code that accesses the database. To help us do that, we will use a test framework that is designed to simplify writing database integration tests, called **DBRider**.

By the end of the chapter, we will have written an integration test against a running database, implemented the fetchesWordByNumber() method from the WordRepository interface, and used the **JDBI** database access library to help us. We will create a database user with permissions on a table storing words to guess. We will create that table, then write a SQL query that JDBI will use to retrieve the word we are looking for. We will use a named parameter SQL query to avoid some application security issues caused by SQL injections.

In this chapter, we're going to cover the following main topics:

- Creating a database integration test
- Implementing the word repository adapter

Technical requirements

The final code for this chapter can be found at https://github.com/PacktPublishing/Test-Driven-Development-with-Java/tree/main/chapter14.

Installing the Postgres database

We will be using the Postgres database in this chapter, which needs installation. To install Postgres, follow these steps:

1. Go to the following web page: https://www.postgresql.org/download/.
2. Follow the installation instructions for your operating system.

    The code has been tested with version 14.5. It is expected to work on all versions.

With the setup completed, let's get started implementing our database code. In the next section, we will use the DBRider framework to create a database integration test.

Creating a database integration test

In this section, we will create the skeleton of a database integration test using a test framework called DBRider. We will use this test to drive out the creation of a database table and database user. We will be working towards implementing the WordRepository interface, which will access words stored in a Postgres database.

Previously, we created a domain model for our Wordz application, using hexagonal architecture to guide us. Instead of accessing a database directly, our domain model uses an abstraction, known as a **port** in hexagonal terminology. One such port is the WordRepository interface, which represents stored words for guessing.

Ports must always be implemented by adapters in hexagonal architecture. An adapter for the WordRepository interface will be a class that implements the interface, containing all the code needed to access the real database.

To test-drive this adapter code, we will write an integration test, using a library that supports testing databases. The library is called DBRider, and is one of the dependencies listed in the project's gradle. build file:

```
dependencies {
    testImplementation 'org.junit.jupiter:junit-jupiter-
api:5.8.2'
    testRuntimeOnly 'org.junit.jupiter:junit-jupiter-
engine:5.8.2'
    testImplementation 'org.assertj:assertj-core:3.22.0'
    testImplementation 'org.mockito:mockito-core:4.8.0'
    testImplementation 'org.mockito:mockito-junit-
jupiter:4.8.0'
    testImplementation 'com.github.database-rider:rider-
core:1.33.0'
    testImplementation 'com.github.database-rider:rider-
junit5:1.33.0'

    implementation 'org.postgresql:postgresql:42.5.0'
}
```

DBRider has an accompanying library called **rider-junit5**, which integrates with **JUnit5**. With this new test tooling, we can start to write our test. The first thing to do is set up the test so that it uses DBRider to connect to our Postgres database.

Creating a database test with DBRider

Before we test-drive any application code, we will need a test that is connected to our Postgres database, running locally. We start in the usual way, by writing a JUnit5 test class:

1. Create a new test class file in the /test/ directory in the new com.wordz.adapters. db package:

Figure 14.1 – Integration test

The IDE will generate the empty test class for us.

2. Add the @DBRider and @DBUnit annotations to the test class:

```
@DBRider
@DBUnit(caseSensitiveTableNames = true,
        caseInsensitiveStrategy= Orthography.LOWERCASE)
public class WordRepositoryPostgresTest {
}
```

The parameters in the @DBUnit annotation mitigate some odd interactions between Postgres and the DBRider test framework to do with case sensitivity on table and column names.

3. We want to test that a word can be fetched. Add an empty test method:

```
@Test
void fetchesWord()  {
}
```

4. Run the test. It will fail:

```
/Users/almellor/Library/Java/JavaVirtualMachines/corretto-17.0.2/Contents/Home/bin/
SLF4J: Failed to load class "org.slf4j.impl.StaticLoggerBinder".
SLF4J: Defaulting to no-operation (NOP) logger implementation
SLF4J: See http://www.slf4j.org/codes.html#StaticLoggerBinder for further details.


java.lang.RuntimeException: JDBC connection url cannot be empty
```

Figure 14.2 – DBRider cannot connect to the database

5. The next step to fixing this is to follow the DBRider documentation and add code that will be used by the DBRider framework. We add a `connectionHolder` field and a `javax.sqlDataSource` field to support that:

```
@DBRider
public class WordRepositoryPostgresTest {
    private DataSource dataSource;


    private final ConnectionHolder connectionHolder
                = () -> dataSource.getConnection();
}
```

The `dataSource` is the standard **JDBC** way of creating a connection to our Postgres database. We run the test. It fails with a different error message:

```
⚠ Tests failed: 1 of 1 test – 95 ms
/Users/almellor/Library/Java/JavaVirtualMachines/corretto-17.0.2/Contents/Home/bin/java ...
SLF4J: Failed to load class "org.slf4j.impl.StaticLoggerBinder".
SLF4J: Defaulting to no-operation (NOP) logger implementation
SLF4J: See http://www.slf4j.org/codes.html#StaticLoggerBinder for further details.

java.lang.NullPointerException: Cannot invoke "javax.sql.DataSource.getConnection()" because "this.dataSource" is null
```

Figure 14.3 – dataSource is null

6. We correct this by adding a `@BeforeEach` method to set up `dataSource`:

```
@BeforeEach
void setupConnection() {
    var ds = new PGSimpleDataSource();
    ds.setServerNames(new String[]{"localhost"});
    ds.setDatabaseName("wordzdb");
    ds.setCurrentSchema("public");
    ds.setUser("ciuser");
```

```
          ds.setPassword("cipassword");


          this.dataSource = ds;
      }
```

This specifies we want a user called `ciuser` with the password `cipassword` to connect to a database called `wordzdb`, running on `localhost` at the default port for Postgres (`5432`).

7.   Run the test and see it fail:

Tests failed: 1 of 1 test – 490 ms

```
java.lang.RuntimeException: Could not initialize database rider datasource.

    at com.github.database.rider.core.connection.RiderDataSource.<init>(RiderDataSource.java:55)
    at com.github.database.rider.core.dataset.DataSetExecutorImpl.getRiderDataSource(DataSetExecutor
    at com.github.database.rider.core.RiderRunner.setup(RiderRunner.java:30)
    at com.github.database.rider.junit5.DBUnitExtension.beforeTestExecution(DBUnitExtension.java:68)
    at java.base/java.util.ArrayList.forEach(ArrayList.java:1511) <9 internal lines>
    at java.base/java.util.ArrayList.forEach(ArrayList.java:1511) <27 internal lines>
Caused by: org.postgresql.util.PSQLException Create breakpoint : FATAL: role "ciuser" does not exist
```

Figure 14.4 – User does not exist

The error is caused because we do not have a `ciuser` user known to our Postgres database yet. Let's create one.

8.   Open a `psql` terminal and create the user:

```
create user ciuser with password 'cipassword';
```

9.   Run the test again:

Tests failed: 1 of 1 test – 446 ms

```
java.lang.RuntimeException: Could not initialize database rider datasource.

    at com.github.database.rider.core.connection.RiderDataSource.<init>(RiderDataSource.java:55)
    at com.github.database.rider.core.dataset.DataSetExecutorImpl.getRiderDataSource(DataSetExecutorIm
    at com.github.database.rider.core.RiderRunner.setup(RiderRunner.java:30)
    at com.github.database.rider.junit5.DBUnitExtension.beforeTestExecution(DBUnitExtension.java:68)
    at java.base/java.util.ArrayList.forEach(ArrayList.java:1511) <9 internal lines>
    at java.base/java.util.ArrayList.forEach(ArrayList.java:1511) <27 internal lines>
Caused by: org.postgresql.util.PSQLException Create breakpoint : FATAL: database "wordzdb" does not exist
```

Figure 14.5 – Database not found

It fails because the DBRider framework is looking to connect our new `ciuser` user to the `wordzdb` database. This database does not exist.

10. In the `psql` terminal, create the database:

```
create database wordzdb;
```

11. Run the test again:

Figure 14.6 – Test passes

The `fetchesWord()` test now passes. We recall that the test method itself is empty, but this means we have enough database set up to proceed with test-driving production code. We will return to database setup soon enough, but we will allow our test-driving to guide us. The next job is to add the missing Arrange, Act, and Assert code to the `fetchesWord()` test.

Driving out the production code

Our goal is to test-drive code to fetch a word from the database. We want that code to be in a class that implements the `WordRepository` interface, which we defined in the domain model. We will want to design enough of our database schema to support this. By starting to add code to the *Assert* step, we can drive out an implementation quickly. This is a useful technique – writing the test by starting with the assert, so that we start with the desired outcome. We can then work backward to include everything necessary for delivering it:

1. Add the Assert step to our `fetchesWord()` test:

```
@Test
public void fetchesWord()  {
    String actual = "";
    assertThat(actual).isEqualTo("ARISE");
}
```

We want to check that we can fetch the word `ARISE` from the database. This test fails. We need to create a class to contain the necessary code.

2. We want our new adapter class to implement the `WordRepository` interface, so we drive this out in the Arrange step of our test:

```
@Test
public void fetchesWord()  {
    WordRepository repository
            = new WordRepositoryPostgres();

    String actual = "";
    assertThat(actual).isEqualTo("ARISE");
}
```

3. We now let the IDE wizard do most of the work in creating our new adapter class. Let's call it `WordRepositoryPostgres`, which links the two facts that the class implements the `WordRepository` interface and is also implementing access to a Postgres database. We use the **New Class** wizard and place it in a new package, `com.wordz.adapters.db`:

Figure 14.7 – New Class wizard

This results in an empty skeleton for the class:

```
package com.wordz.adapters.db;

import com.wordz.domain.WordRepository;

public class WordRepositoryPostgres implements
                                WordRepository {
}
```

4. The IDE will auto-generate method stubs for the interface:

```java
public class WordRepositoryPostgres implements
WordRepository {
    @Override
    public String fetchWordByNumber(int number) {
        return null;
    }

    @Override
    public int highestWordNumber() {
        return 0;
    }
}
```

5. Returning to our test, we can add the act line, which will call the fetchWordByNumber() method:

```java
@Test
public void fetchesWord()  {
    WordRepository repository
                = new WordRepositoryPostgres();

    String actual =
            repository.fetchWordByNumber(27);

    assertThat(actual).isEqualTo("ARISE");
}
```

A word of explanation about the mysterious constant 27 passed in to the fetchWordByNumber() method. This is an *arbitrary* number used to identify a particular word. Its only hard requirement is that it must line up with the word number given in the stub test data, which we will see a little later in a JSON file. The actual value of 27 is of no significance beyond lining up with the word number of the stub data.

6. Pass dataSource in to the WordRepositoryPostgres constructor so that our class has a way to access the database:

```java
@Test
public void fetchesWord()  {
    WordRepository repository
            = new
```

```
                  WordRepositoryPostgres(dataSource);

            String actual = adapter.fetchWordByNumber(27);

            assertThat(actual).isEqualTo("ARISE");
        }
```

This drives out a change to the constructor:

```
        public WordRepositoryPostgres(DataSource dataSource){
            // Not implemented
        }
```

7. The last bit of setup to do in our test is to populate the database with the word ARISE. We do this using a JSON file that the DBRider framework will apply to our database on test startup:

```
    {
        "word": [
            {
                "word_number": 27,
                "word": "ARISE"
            }
        ]
    }
```

The "word_number": 27 code here corresponds to the value used in the test code.

8. This file must be saved in a specific location so that DBRider can find it. We call the file wordTable.json and save it in the test directory, in /resources/adapters/data:

Figure 14.8 – Location of wordTable.json

9.   The final step in setting up our failing test is to link the test data `wordTable.json` file to our `fetchesWord()` test method. We do this using the DBRider `@DataSet` annotation:

```
@Test
@DataSet("adapters/data/wordTable.json")
public void fetchesWord()  {
    WordRepository repository
        = new WordRepositoryPostgres(dataSource);

    String actual =
                repository.fetchWordByNumber(27);

    assertThat(actual).isEqualTo("ARISE");
}
```

The test now fails and is in a position where we can make it pass by writing the database access code. In the next section, we will use the popular library JDBI to implement database access in an adapter class for our `WordRepository` interface.

Implementing the WordRepository adapter

In this section, we will use the popular database library JDBI to implement the `fetchWordByNumber()` method of `interface WordRepository` and make our failing integration test pass.

Hexagonal architectures were covered in *Chapter 9, Hexagonal Architecture – Decoupling External Systems*. An external system like a database is accessed through a port in the domain model. The code that is specific to that external system is contained in an adapter. Our failing test enables us to write the database access code to fetch a word to guess.

A little bit of database design thinking needs to be done before we begin writing code. For the task at hand, it is enough to note that we will store all available words to guess in a database table named `word`. This table will have two columns. There will be a primary key named `word_number` and a five-letter word in a column named `word`.

Let's test-drive this out:

1.   Run the test to reveal that the `word` table does not exist:

Tests failed: 1 of 1 test – 797 ms

```
l.search.SearchException Create breakpoint : org.dbunit.dataset.NoSuchTableException: The table 'word' does not exist in schema 'null'
e.search.AbstractMetaDataBasedSearchCallback.getNodes(AbstractMetaDataBasedSearchCallback.java:154)
```

Figure 14.9 – Table not found

2. Correct this by creating a `word` table in the database. We use the `psql` console to run the SQL `create table` command:

```
create table word (word_number int primary key,
word char(5));
```

3. Run the test again. The error changes to show our `ciuser` user has insufficient permissions:

```
❶ Tests failed: 1 of 1 test – 1 sec 8 ms

Caused by: org.postgresql.util.PSQLException Create breakpoint : ERROR: permission denied for table word
    at org.postgresql.core.v3.QueryExecutorImpl.receiveErrorResponse(QueryExecutorImpl.java:2674)
```

Figure 14.10 – Insufficient permissions

4. We correct this by running the SQL `grant` command in the `psql` console:

```
grant select, insert, update, delete on all tables in
schema public to ciuser;
```

5. Run the test again. The error changes to show us that the `word` has not been read from the database table:

Figure 14.11 – Word not found

Accessing the database

Having set up the database side of things, we can move on to adding the code that will access the database. The first step is to add the database library we will use. It is JDBI, and to use it, we must add the `jdbi3-core` dependency to our `gradle.build` file:

```
dependencies {
    testImplementation 'org.junit.jupiter:junit-jupiter-
api:5.8.2'
    testRuntimeOnly 'org.junit.jupiter:junit-jupiter-
```

```
engine:5.8.2'
    testImplementation 'org.assertj:assertj-core:3.22.0'
    testImplementation 'org.mockito:mockito-core:4.8.0'
    testImplementation 'org.mockito:mockito-junit-
jupiter:4.8.0'
    testImplementation 'com.github.database-rider:rider-
core:1.35.0'
    testImplementation 'com.github.database-rider:rider-
junit5:1.35.0'

    implementation 'org.postgresql:postgresql:42.5.0'
    implementation 'org.jdbi:jdbi3-core:3.34.0'
}
```

> **Note**
> The code itself is as described in the JDBI documentation, found here: `https://jdbi.org/#_queries`.

Follow these steps to access the database:

1. First, create a `jdbi` object in the constructor of our class:

```
public class WordRepositoryPostgres
                        implements WordRepository {
    private final Jdbi jdbi;

    public WordRepositoryPostgres(DataSource
                                        dataSource){
        jdbi = Jdbi.create(dataSource);
    }
}
```

This gives us access to the JDBI library. We have arranged it so that JDBI will access whatever `DataSource` we pass into our constructor.

2. We add the JDBI code to send a SQL query to the database and fetch the word corresponding to the `wordNumber` we provide as a method parameter. First, we add the SQL query we will use:

```
private static final String SQL_FETCH_WORD_BY_NUMBER
    = "select word from word where "
                    + "word_number=:wordNumber";
```

3. The `jdbi` access code can be added to the `fetchWordByNumber()` method:

```
@Override
public String fetchWordByNumber(int wordNumber) {
    String word = jdbi.withHandle(handle -> {
        var query =
          handle.createQuery(SQL_FETCH_WORD_BY_NUMBER);
        query.bind("wordNumber", wordNumber);

        return query.mapTo(String.class).one();
    });

    return word;
}
```

4. Run the test again:

Figure 14.12 – Test passing

Our integration test now passes. The adapter class has read the word from the database and returned it.

Implementing GameRepository

The same process is used to test-drive the `highestWordNumber()` method and to create adapters for the other database access code implementing the `GameRepository` interface. The final code for these can be seen on GitHub with comments to explore some of the issues in database testing, such as how to avoid test failures caused by stored data.

There is a manual step needed to test-drive the implementation code for the `GameRepository` interface. We must create a game table.

In psql, type the following:

```
CREATE TABLE game (
    player_name character varying NOT NULL,
    word character(5),
    attempt_number integer DEFAULT 0,
    is_game_over boolean DEFAULT false
);
```

Summary

In this chapter, we have created an integration test for our database. We used that to test-drive the implementation of a database user, the database table, and the code needed to access our data. This code implemented the adapter for one of our ports in our hexagonal architecture. Along the way, we used some new tools. The DBRider database test framework simplified our test code. The JDBI database access library simplified our data access code.

In the next and final chapter, *Chapter 15, Driving the Web Layer*, we will add an HTTP interface to our application, turning it into a complete microservice. We will integrate all the components together, then play our first game of Wordz using the HTTP test tool Postman.

Questions and answers

1. Should we automate the manual steps of creating the database?

    Yes. This is an important part of DevOps, where we developers are responsible for getting the code into production and keeping it running there. The key technique is **Infrastructure as Code (IaC)**, which means automating manual steps as code that we check in to the main repository.

2. What tools can help with automating database creation?

    Popular tools are **Flyway** and **Liquibase**. Both allow us to write scripts that are run at application startup and will migrate the database schema from one version to the next. They assist in migrating data across schema changes where that is required. These are outside the scope of this book.

3. What tools can help with installing the database?

    Access to a running database server is part of platform engineering. For cloud-native designs that run on Amazon Web Service, Microsoft Azure, or Google Cloud Platform, use configuration scripting for that platform. One popular approach is to use Hashicorp's **Terraform**, which aims to be a cross-provider universal scripting language for cloud configuration. This is outside of the scope of this book.

4. How often should we run the integration tests?

Before every check-in to the repository. While unit tests are fast to run and should be run all the time, integration tests by nature are slower to execute. It is reasonable to run only unit tests while working on domain code. We must always ensure we haven't broken anything unexpectedly. This is where running integration tests comes in. These reveal whether we have accidentally changed something that affects the adapter layer code, or whether something has changed regarding database layout.

Further reading

- Documentation for DBRider: `https://github.com/database-rider/database-rider`

- JDBI documentation: `https://jdbi.org/#_introduction_to_jdbi_3`

- Flyway is a library that allows us to store the SQL commands to create and modify our database schema as source code. This allows us to automate database changes: `https://flywaydb.org/`

- As our application design grows, our database schema will need to change. This website and the accompanying books describe ways to do this while managing risk: `https://databaserefactoring.com/`

- Hosting a Postgres database on Amazon Web Services using their RDS service: `https://aws.amazon.com/rds`

15
Driving the Web Layer

In this chapter, we complete our web application by adding a web endpoint. We will learn how to write HTTP integration tests using the built-in Java HTTP client. We will test-drive the web adapter code that runs this endpoint, using an open source HTTP server framework. This web adapter is responsible for converting HTTP requests into commands we can execute in our domain layer. At the end of the chapter, we will assemble all the pieces of our application into a microservice. The web adapter and database adapters will be linked to the domain model using dependency injection. We will need to run a few manual database commands, install a web client called Postman, and then we can play our game.

In this chapter, we're going to cover the following main topics:

- Starting a new game
- Playing the game
- Integrating the application
- Using the application

Technical requirements

The code for this chapter is available at `https://github.com/PacktPublishing/Test-Driven-Development-with-Java/tree/main/chapter15`.

Before attempting to run the final application, perform the following steps:

1. Ensure the **Postgres** database is running locally.
2. Ensure the database setup steps from *Chapter 14 , Driving the Database Layer,* have been completed.

3. Open the **Postgres pqsl** command terminal and enter the following SQL command:

```
insert into word values (1, 'ARISE'), (2, 'SHINE'), (3,
'LIGHT'), (4, 'SLEEP'), (5, 'BEARS'), (6, 'GREET'), (7,
'GRATE');
```

4. Install **Postman** by following the instructions at https://www.postman.com/downloads/.

Starting a new game

In this section, we will test-drive a web adapter that will provide our domain model with an HTTP API. External web clients will be able to send HTTP requests to this endpoint to trigger actions in our domain model so that we can play the game. The API will return appropriate HTTP responses, indicating the score for the submitted guess and reporting when the game is over.

The following open source libraries will be used to help us write the code:

- Molecule: This is a lightweight HTTP framework
- Undertow: This is a lightweight HTTP web server that powers the Molecule framework
- GSON: This is a Google library that converts between Java objects and JSON structured data

To start building, we first add the required libraries as dependencies to the build.gradle file. Then we can begin writing an integration test for our HTTP endpoint and test-drive the implementation.

Adding required libraries to the project

We need to add the three libraries Molecule, Undertow, and Gson to the build.gradle file before we can use them:

Add the following code to the build.gradle file:

```
dependencies {
    testImplementation 'org.junit.jupiter:junit-jupiter-
api:5.8.2'
    testRuntimeOnly 'org.junit.jupiter:junit-jupiter-
engine:5.8.2'
    testImplementation 'org.assertj:assertj-core:3.22.0'
    testImplementation 'org.mockito:mockito-core:4.8.0'
    testImplementation 'org.mockito:mockito-junit-
jupiter:4.8.0'
    testImplementation 'com.github.database-rider:rider-
core:1.35.0'
```

```
    testImplementation 'com.github.database-rider:rider-
junit5:1.35.0'

    implementation 'org.postgresql:postgresql:42.5.0'
    implementation 'org.jdbi:jdbi3-core:3.34.0'
    implementation 'org.apache.commons:commons-lang3:3.12.0'
    implementation 'com.vtence.molecule:molecule:0.15.0'
    implementation 'io.thorntail:undertow:2.7.0.Final'
    implementation 'com.google.code.gson:gson:2.10'
}
```

Writing the failing test

We will follow the normal TDD cycle to create our web adapter. When writing tests for objects in the adapter layer, we must focus on testing the translation between objects in our domain layer and communications with external systems. Our adapter layer will use the Molecule HTTP framework to handle HTTP requests and responses.

As we have used hexagonal architecture and started with the domain layer, we already know that the game logic is working. The goal of this test is to prove that the web adapter layer is performing its responsibility. That is to translate HTTP requests and responses to objects in our domain layer.

As ever, we begin by creating a test class:

1. First, we write our test class. We'll call it `WordzEndpointTest`, and it belongs in the `com.wordz.adapters.api` package:

    ```
    package com.wordz.adapters.api;


    public class WordzEndpointTest {
    }
    ```

    The reason for including this package is as part of our hexagonal architecture. Code in this web adapter is allowed to use anything from the domain model. The domain model itself is unaware of the existence of this web adapter.

    Our first test will be to start a new game:

    ```
    @Test
    void startGame() {
    }
    ```

2. This test needs to capture the design decision that surrounds our intended web API. One decision is that when a game has successfully started, we will return a simple 204 No Content HTTP status code. We will start with the assert to capture this decision:

```
@Test
void startGame() {
    HttpResponse res;
    assertThat(res)
        .hasStatusCode(HttpStatus.NO_CONTENT.code);
}
```

3. Next, we write the Act step. The action here is for an external HTTP client to send a request to our web endpoint. To achieve this, we use the built-in HTTP client provided by Java itself. We arrange the code to send the request, and then discard any HTTP response body, as our design does not return a body:

```
@Test
void startGame() throws IOException,
                        InterruptedException {

    var httpClient = HttpClient.newHttpClient();
    HttpResponse res
        = httpClient.send(req,
            HttpResponse.BodyHandlers.discarding());

    assertThat(res)
        .hasStatusCode(HttpStatus.NO_CONTENT.code);
}
```

4. The Arrange step is where we capture our decisions about the HTTP request to send. In order to start a new game, we need a Player object to identify the player. We will send this as a Json object in the Request body. The request will cause a state change on our server, so we choose the HTTP POST method to represent that. Finally, we choose a route whose path is /start:

```
@Test
private static final Player PLAYER
        = new Player("alan2112");

void startGame() throws IOException,
                        InterruptedException {
```

```
        var req = HttpRequest.newBuilder()
          .uri(URI.create("htp://localhost:8080/start"))
          .POST(HttpRequest.BodyPublishers
                .ofString(new Gson().toJson(PLAYER)))
                .build();

        var httpClient = HttpClient.newHttpClient();
        HttpResponse res
            = httpClient.send(req,
                HttpResponse.BodyHandlers.discarding());

        assertThat(res)
           .hasStatusCode(HttpStatus.NO_CONTENT.code);
    }
```

We see the Gson library being used to convert a `Player` object into its JSON representation. We also see a `POST` method is constructed and sent to the `/start` path on `localhost`. Eventually, we will want to move the `localhost` detail into configuration. But, for now, it will get the test working on our local machine.

5. We can run our integration test and confirm that it fails:

Figure 15.1 – A failed test – no HTTP server

Unsurprisingly, this test fails because it cannot connect to an HTTP server. Fixing that is our next task.

Creating our HTTP server

The failing test allows us to test-drive code that implements an HTTP server. We will use the Molecule library to provide HTTP services to us:

1. Add an endpoint class, which we will call `class WordzEndpoint`:

    ```
    @Test
    void startGame() throws IOException,
                            InterruptedException {
        var endpoint
            = new WordzEndpoint("localhost", 8080);
    ```

    The two parameters passed into the `WordzEndpoint` constructor define the host and port that the web endpoint will run on.

2. Using the IDE, we generate the class:

    ```
    package com.wordz.adapters.api;


    public class WordzEndpoint {
        public WordzEndpoint(String host, int port) {
        }
    }
    ```

    In this case, we're not going to store the host and port details in fields. Instead, we are going to start a `WebServer` using a class from the Molecule library.

3. Create a `WebServer` using the Molecule library:

    ```
    package com.wordz.adapters.api;


    import com.vtence.molecule.WebServer;

    public class WordzEndpoint {
        private final WebServer server;

        public WordzEndpoint(String host, int port) {
            server = WebServer.create(host, port);
        }
    }
    ```

The preceding code is enough to start an HTTP server running and allow the test to connect to it. Our HTTP server does nothing useful in terms of playing our game. We need to add some routes to this server along with the code to respond to them.

Adding routes to the HTTP server

To be useful, the HTTP endpoint must respond to HTTP commands, interpret them, and send them as commands to our domain layer. As design decisions, we decide on the following:

- That a /start route must be called to start the game
- That we will use the HTTP POST method
- That we will identify which player the game belongs to as JSON data in the POST body

To add routes to the HTTP server, do the following:

1. Test-drive the /start route. To work in small steps, initially, we will return a NOT_
   IMPLEMENTED HTTP response code:

```
public class WordzEndpoint {
    private final WebServer server;

    public WordzEndpoint(String host, int port) {
        server = WebServer.create(host, port);

        try {
            server.route(new Routes() {{
                post("/start")
                    .to(request -> startGame(request));
            }});
        } catch (IOException ioe) {
            throw new IllegaStateException(ioe);
        }
    }

    private Response startGame(Request request) {
        return Response
                .of(HttpStatus.NOT_IMPLEMENTED)
                .done();
```

```
    }
}
```

2. We can run the `WordzEndpointTest` integration test:

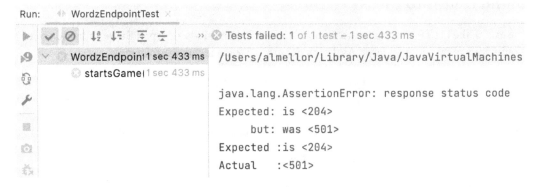

Figure 15.2 – An incorrect HTTP status

The test fails, as expected. We have made progress because the test now fails for a different reason. We can now connect to the web endpoint, but it does not return the right HTTP response. Our next task is to connect this web endpoint to the domain layer code and take the relevant actions to start a game.

Connecting to the domain layer

Our next task is to receive an HTTP request and translate that into domain layer calls. This involves parsing JSON request data, using the Google Gson library, into Java objects, then sending that response data to the `class Wordz` port:

1. Add the code to call the domain layer port implemented as `class Wordz`. We will use `Mockito` to create a test double for this object. This allows us to test only the web endpoint code, decoupled from all other code:

```
@ExtendWith(MockitoExtension.class)
public class WordzEndpointTest {
    @Mock
    private Wordz mockWordz;

    @Test
    void startGame() throws IOException,
                            InterruptedException {
        var endpoint
```

```
                    = new WordzEndpoint(mockWordz,
                                    "localhost", 8080);
```

2. We need to provide our `class Wordz` domain object to the `class WordzEndpoint` object. We use dependency injection to inject it into the constructor:

```
public class WordzEndpoint {
    private final WebServer server;
    private final Wordz wordz;


    public WordzEndpoint(Wordz wordz,
                        String host, int port) {
        this.wordz = wordz;
```

3. Next, we need to add the code to start a game. To do that, we first extract the `Player` object from the JSON data in the `request` body. That identifies which player to start a game for. Then we call the `wordz.newGame()` method. If it is successful, we return an HTTP status code of `204 No Content`, indicating success:

```
private Response startGame(Request request) {
    try {
        Player player
                = new Gson().fromJson(request.body(),
                                        Player.class);


        boolean isSuccessful = wordz.newGame(player);
        if (isSuccessful) {
            return Response
                    .of(HttpStatus.NO_CONTENT)
                    .done();
        }
    } catch (IOException e) {
        throw new RuntimeException(e);
    }

    throw new
        UnsupportedOperationException("Not
                                implemented");
}
```

4. Now, we can run the test, however, it fails:

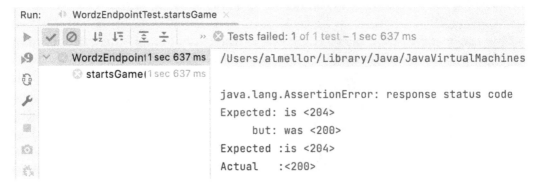

Figure 15.3 – An incorrect HTTP response

It failed because the return value from wordz.newGame() was false. The mock object needs to be set up to return true.

5. Return the correct value from the mockWordz stub:

```
    @Test
  void startsGame() throws IOException,
                        InterruptedException {
    var endpoint
        = new WordzEndpoint(mockWordz,
                            "localhost", 8080);

    when(mockWordz.newGame(eq(PLAYER)))
        .thenReturn(true);
```

6. Then, run the test:

Figure 15.4 – The test passes

The integration test passes. The HTTP request has been received, called the domain layer code to start a new game, and the HTTP response is returned. The next step is to consider refactoring.

Refactoring the start game code

As usual, once a test passes, we consider what – if anything – we need to refactor.

It will be worthwhile to refactor the test to simplify writing new tests by collating common code into one place:

```java
@ExtendWith(MockitoExtension.class)
public class WordzEndpointTest {
    @Mock
    private Wordz mockWordz;

    private WordzEndpoint endpoint;

    private static final Player PLAYER
                        = new Player("alan2112");

    private final HttpClient httpClient
                        = HttpClient.newHttpClient();

    @BeforeEach
    void setUp() {
        endpoint = new WordzEndpoint(mockWordz,
                                "localhost", 8080);
    }

    @Test
    void startsGame() throws IOException,
                             InterruptedException {
        when(mockWordz.newGame(eq(player)))
                        .thenReturn(true);

        var req = requestBuilder("start")
                .POST(asJsonBody(PLAYER))
                .build();
```

```
        var res
          = httpClient.send(req,
                HttpResponse.BodyHandlers.discarding());

        assertThat(res)
            .hasStatusCode(HttpStatus.NO_CONTENT.code);
    }

    private HttpRequest.Builder requestBuilder(
        String path) {
        return HttpRequest.newBuilder()
                .uri(URI.create("http://localhost:8080/"
                                + path));
    }

    private HttpRequest.BodyPublisher asJsonBody(
        Object source) {
        return HttpRequest.BodyPublishers
                .ofString(new Gson().toJson(source));
    }
}
```

Handling errors when starting a game

One of our design decisions is that a player cannot start a game when one is in progress. We need to test-drive this behavior. We choose to return an HTTP status of 409 Conflict to indicate that a game is already in progress for a player and a new one cannot be started for them:

1. Write the test to return a 409 Conflict if the game is already in progress:

    ```
    @Test
    void rejectsRestart() throws Exception {
        when(mockWordz.newGame(eq(player)))
                        .thenReturn(false);

        var req = requestBuilder("start")
                .POST(asJsonBody(player))
    ```

```
                            .build();

            var res
                = httpClient.send(req,
                    HttpResponse.BodyHandlers.discarding());

            assertThat(res)
                    .hasStatusCode(HttpStatus.CONFLICT.code);
        }
```

2. Next, run the test. It should fail, as we have yet to write the implementation code:

Figure 15.5 – A failing test

3. Test-drive the code to report that the game cannot be restarted:

```
    private Response startGame(Request request) {
        try {
            Player player
                    = new Gson().fromJson(request.body(),
                                            Player.class);

            boolean isSuccessful = wordz.newGame(player);
            if (isSuccessful) {
                return Response
                        .of(HttpStatus.NO_CONTENT)
                        .done();
            }
```

```
        return Response
               .of(HttpStatus.CONFLICT)
               .done();
    } catch (IOException e) {
        throw new RuntimeException(e);
    }
}
```

4.   Run the test again:

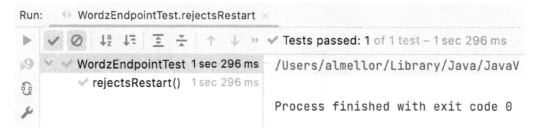

Figure 15. 6 – The test passes

The test passes when run on its own now that the implementation is in place. Let's run all the
WordzEndpointTests tests to double-check our progress.

5.   Run all WordzEndpointTests:

Figure 15.7 – Test failure due to restarting the server

Unexpectedly, the tests fail when run one after the other.

Fixing the unexpectedly failing tests

When we run all of the tests, they now fail. The tests all previously ran correctly when run one at a
time. A recent change has clearly broken something. We lost our test isolation at some point. This
error message indicates the web server is being started twice on the same port, which is not possible.

The options are to stop the web server after each test or to only start the web server once for all tests. As this is intended to be a long-running microservice, only starting once seems the better choice here:

1. Add a @BeforeAll annotation to only start the HTTP server once:

```
@BeforeAll
void setUp() {
    mockWordz = mock(Wordz.class);
    endpoint = new WordzEndpoint(mockWordz,
                                 "localhost", 8080);
}
```

We change the @BeforeEach annotation to a @BeforeAll annotation to make the endpoint creation only happen once per test. To support this, we also must create the mock and use an annotation on the test itself to control the life cycle of objects:

```
@ExtendWith(MockitoExtension.class)
@TestInstance(TestInstance.Lifecycle.PER_CLASS)
public class WordzEndpointTest {
```

Both tests in WordzEndpointTest now pass.

2. With all tests passing again, we can consider refactoring the code. A readability improvement will come from extracting an extractPlayer() method. We can also make the conditional HTTP status code more concise:

```
private Response startGame(Request request) {
    try {
        Player player = extractPlayer(request);

        boolean isSuccessful = wordz.newGame(player);
        HttpStatus status
                = isSuccessful?
                    HttpStatus.NO_CONTENT :
                    HttpStatus.CONFLICT;

        return Response
                    .of(status)
                    .done();
    } catch (IOException e) {
        throw new RuntimeException(e);
```

```
        }
    }


    private Player extractPlayer(Request request)
                                    throws IOException {
        return new Gson().fromJson(request.body(),
                                    Player.class);
    }
}
```

We have now completed the major part of the coding needed to start a game. To handle the remaining error condition, we can now test-drive the code to return 400 BAD REQUEST if the Player object cannot be read from the JSON payload. We will omit that code here. In the next section, we will move on to test-driving the code for guessing the target word.

Playing the game

In this section, we will test-drive the code to play the game. This involves submitting multiple guess attempts to the endpoint until a game-over response is received.

We start by creating an integration test for the new /guess route in our endpoint:

1. The first step is to code the Arrange step. Our domain model provides the assess() method on class Wordz to assess the score for a guess, along with reporting whether the game is over. To test-drive this, we set up the mockWordz stub to return a valid GuessResult object when the assess() method is called:

    ```
    @Test
    void partiallyCorrectGuess() {
        var score = new Score("-U---");
        score.assess("GUESS");
        var result = new GuessResult(score, false, false);
        when(mockWordz.assess(eq(player), eq("GUESS")))
                .thenReturn(result);
    }
    ```

2. The Act step will call our endpoint with a web request submitting the guess. Our design decision is to send an HTTP POST request to the /guess route. The request body will contain a JSON representation of the guessed word. To create this, we will use record GuessRequest and use Gson to convert that into JSON for us:

    ```
    @Test
    void partiallyCorrectGuess() {
    ```

```
        var score = new Score("-U---");
        score.assess("GUESS");
        var result = new GuessResult(score, false, false);
        when(mockWordz.assess(eq(player), eq("GUESS")))
                .thenReturn(result);

        var guessRequest = new GuessRequest(player, "-U---");
        var body = new Gson().toJson(guessRequest);
        var req = requestBuilder("guess")
                .POST(ofString(body))
                .build();
}
```

3. Next, we define the record:

```
package com.wordz.adapters.api;

import com.wordz.domain.Player;

public record GuessRequest(Player player, String guess) {
}
```

4. Then, we send the request over HTTP to our endpoint, awaiting the response:

```
@Test
void partiallyCorrectGuess() throws Exception {
    var score = new Score("-U---");
    score.assess("GUESS");
    var result = new GuessResult(score, false, false);
    when(mockWordz.assess(eq(player), eq("GUESS")))
            .thenReturn(result);

    var guessRequest = new GuessRequest(player, "-U---");
    var body = new Gson().toJson(guessRequest);
    var req = requestBuilder("guess")
            .POST(ofString(body))
            .build();
```

```
        var res
          = httpClient.send(req,
                HttpResponse.BodyHandlers.ofString());
    }
```

5. Then, we extract the returned body data and assert it against our expectations:

```
@Test
void partiallyCorrectGuess() throws Exception {
    var score = new Score("-U--G");
    score.assess("GUESS");
    var result = new GuessResult(score, false, false);
    when(mockWordz.assess(eq(player), eq("GUESS")))
            .thenReturn(result);


    var guessRequest = new GuessRequest(player,
                                        "-U--G");
    var body = new Gson().toJson(guessRequest);
    var req = requestBuilder("guess")
            .POST(ofString(body))
            .build();

    var res
        = httpClient.send(req,
            HttpResponse.BodyHandlers.ofString());

    var response
        = new Gson().fromJson(res.body(),
                        GuessHttpResponse.class);

    // Key to letters in scores():
    // C correct, P part correct, X incorrect
    Assertions.assertThat(response.scores())
        .isEqualTo("PCXXX");
    Assertions.assertThat(response.isGameOver())
```

```
            .isFalse();
    }
```

One API design decision here is to return the per-letter scores as a five-character `String` object. The single letters X, C, and P are used to indicate incorrect, correct, and partially correct letters. We capture this decision in the assertion.

6. We define a record to represent the JSON data structure we will return as a response from our endpoint:

```
package com.wordz.adapters.api;


public record GuessHttpResponse(String scores,
                                boolean isGameOver) {

}
```

7. As we have decided to POST to a new /guess route, we need to add this route to the routing table. We also need to bind it to a method that will take action, which we will call guessWord():

```
public WordzEndpoint(Wordz wordz, String host,
                     int port) {
    this.wordz = wordz;
    server = WebServer.create(host, port);

    try {
        server.route(new Routes() {{
            post("/start")
                .to(request -> startGame(request));
            post("/guess")
                .to(request -> guessWord(request));
        }});
    } catch (IOException e) {
        throw new IllegalStateException(e);
    }
}
```

We add an `IllegalStateException` to rethrow any problems that occur when starting the HTTP server. For this application, this exception may propagate upwards and cause the application to stop running. Without a working web server, none of the web code makes sense to run.

8. We implement the guessWord() method with code to extract the request data from the POST body:

```
private Response guessWord(Request request) {
    try {
        GuessRequest gr =
            extractGuessRequest(request);
        return null ;
    } catch (IOException e) {
        throw new RuntimeException(e);
    }
}
private GuessRequest extractGuessRequest(Request request)
throws IOException {
    return new Gson().fromJson(request.body(),
                            GuessRequest.class);
}
```

9. Now we have the request data, it's time to call our domain layer to do the real work. We will capture the GuessResult object returned, so we can base our HTTP response from the endpoint on it:

```
private Response guessWord(Request request) {
    try {
        GuessRequest gr =
            extractGuessRequest(request);
        GuessResult result
            = wordz.assess(gr.player(),
                gr.guess());
        return null;
    } catch (IOException e) {
        throw new RuntimeException(e);
    }
}
```

10. We choose to return a different format of data from our endpoint compared to the GuessResult object returned from our domain model. We will need to transform the result from the domain model:

```
private Response guessWord(Request request) {
    try {
        GuessRequest gr =
            extractGuessRequest(request);
        GuessResult result = wordz.assess(gr.player(),
                                gr.guess());
        return Response.ok()
                .body(createGuessHttpResponse(result))
                .done();
    } catch (IOException e) {
        throw new RuntimeException(e);
    }
}


private String createGuessHttpResponse(GuessResult
result) {
    GuessHttpResponse httpResponse
        = new
            GuessHttpResponseMapper().from(result);
    return new Gson().toJson(httpResponse);
}
```

11. We add an empty version of the object doing the transformation, which is class GuessHttpResponseMapper. In this first step, it will simply return null:

```
package com.wordz.adapters.api;

import com.wordz.domain.GuessResult;

public class GuessHttpResponseMapper {
    public GuessHttpResponse from(GuessResult result) {
        return null;
    }
}
```

12. This is enough to compile and be able to run the `WordzEndpointTest` test:

Figure 15.8 – The test fails

13. With a failing test in place, we can now test-drive the details of the transform class. To do this, we switch to adding a new unit test called `class GuessHttpResponseMapperTest`.

> **Note**
>
> The details of this are omitted but can be found on GitHub – it follows the standard approach used throughout the book.

14. Once we have test-driven the detailed implementation of `class GuessHttpResponseMapper`, we can rerun the integration test:

Figure 15.9 – The endpoint test passes

As we see in the preceding image, the integration test has passed! Time for a well-earned coffee break. Well, mine's a nice English breakfast tea, but that's just me. After that, we can test-drive the response to any errors that occurred. Then it's time to bring the microservice together. In the next section, we will assemble our application into a running microservice.

Integrating the application

In this section, we will bring together the components of our test-driven application. We will form a microservice that runs our endpoint and provides the frontend web interface to our service. It will use the Postgres database for storage.

We need to write a short main() method to link together the major components of our code. This will involve creating concrete objects and injecting dependencies into constructors. The main() method exists on class WordzApplication, which is the entry point to our fully integrated web service:

```
package com.wordz;

import com.wordz.adapters.api.WordzEndpoint;
import com.wordz.adapters.db.GameRepositoryPostgres;
import com.wordz.adapters.db.WordRepositoryPostgres;
import com.wordz.domain.Wordz;

public class WordzApplication {
    public static void main(String[] args) {
        var config = new WordzConfiguration(args);
        new WordzApplication().run(config);
    }

    private void run(WordzConfiguration config) {
        var gameRepository
          = new GameRepositoryPostgres(config.getDataSource());

        var wordRepository
          = new WordRepositoryPostgres(config.getDataSource());

        var randomNumbers = new ProductionRandomNumbers();

        var wordz = new Wordz(gameRepository,
                              wordRepository,
                              randomNumbers);

        var api = new WordzEndpoint(wordz,
                                    config.getEndpointHost(),
                                    config.getEndpointPort());

        waitUntilTerminated();
    }
```

```
private void waitUntilTerminated() {
    try {
        while (true) {
            Thread.sleep(10000);
        }
    } catch (InterruptedException e) {
        return;
    }
}
}
```

The main() method instantiates the domain model, and dependency injects the concrete version of our adapter classes into it. One notable detail is the waitUntilTerminated() method. This prevents main() from terminating until the application is closed down. This, in turn, keeps the HTTP endpoint responding to requests.

Configuration data for the application is held in class WordzConfiguration. This has default settings for the endpoint host and port settings, along with database connection settings. These can also be passed in as command line arguments. The class and its associated test can be seen in the GitHub code for this chapter.

In the next section, we will use the Wordz web service application using the popular HTTP testing tool, Postman.

Using the application

To use our newly assembled web application, first ensure that the database setup steps and the Postman installation described in the *Technical requirements* section have been successfully completed. Then run the main() method of class WordzApplication in IntelliJ. That starts the endpoint, ready to accept requests.

Once the service is running, the way we interact with it is by sending HTTP requests to the endpoint. Launch Postman and (on macOS) a window that looks like this will appear:

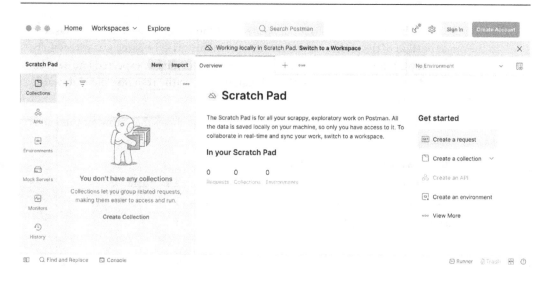

Figure 15.10 – Postman home screen

We first need to start a game. To do that, we need to send HTTP POST requests to the /start route on our endpoint. By default, this will be available at http://localhost:8080/start. We need to send a body, containing the JSON {"name":"testuser"} text.

We can send this request from Postman. We click the **Create a request** button on the home page. This takes us to a view where we can enter the URL, select the **POST** method and type our JSON body data:

1. Create a **POST** request to start the game:

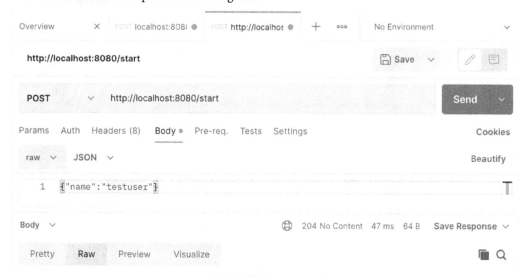

Figure 15.11 – Start a new game

Click the blue **Send** button. The screenshot in *Figure 15.11* shows both the request that was sent – in the upper portion of the screen – and the response. In this case, the game was successfully started for the player named `testuser`. The endpoint performed as expected and sent an HTTP status code of `204 No Content`. This can be seen in the response panel, towards the bottom of the screenshot.

A quick check of the contents of the `game` table in the database shows that a row for this game has been created:

```
wordzdb=# select * from game;
 player_name | word  | attempt_number | is_game_over
-------------+-------+----------------+--------------
 testuser    | ARISE |              0 | f
(1 row)


wordzdb=#
```

2. We can now make our first guess at the word. Let's try a guess of `"STARE"`. The **POST** request for this and the response from our endpoint appears, as shown in the following screenshot:

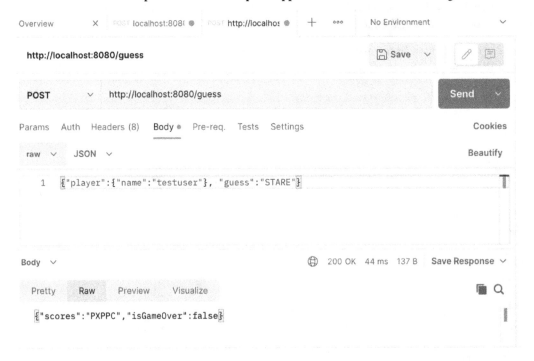

Figure 15.12 – Score returned

The endpoint returns an HTTP status code of 200 OK. This time, a body of JSON formatted data is returned. We see "scores":"PXPPC" indicating that the first letter of our guess, S, appears in the word somewhere but not in the first position. The second letter of our guess, T, is incorrect and does not appear in the target word. We got two more part-correct letters and one final correct letter in our guess, which was the letter E at the end.

The response also shows "isGameOver":false. We haven't finished the game yet.

3. We will make one more guess, cheating slightly. Let's send a **POST** request with a guess of "ARISE":

Figure 15.13 – A successful guess

Winner! We see "scores":"CCCCC" telling us all five letters of our guess are correct. "isGameOver":true tells us that our game has ended, on this occasion, successfully.

We've successfully played one game of Wordz using our microservice.

Summary

In this section, we have completed our Wordz application. We used an integration test with TDD to drive out an HTTP endpoint for Wordz. We used open source HTTP libraries – Molecule, Gson, and Undertow. We made effective use of hexagonal architecture. Using ports and adapters, these frameworks became an implementation detail rather than a defining feature of our design.

We assembled our final application to bring together the business logic held in the domain layer with the Postgres database adapter and the HTTP endpoint adapter. Working together, our application forms a small microservice.

In this final chapter, we have arrived at a small-scale yet typical microservice comprising an HTTP API and a SQL database. We've developed the code test first, using tests to guide our design choices. We have applied the SOLID principles to improve how our software fits together. We have learned how the ports and adapters of hexagonal architecture simplify the design of code that works with external systems. Using hexagonal architecture is a natural fit for TDD, allowing us to develop our core application logic with FIRST unit tests. We have created both a database adapter and an HTTP adapter test first, using integration tests. We applied the rhythms of TDD – Red, Green, Refactor and Arrange, Act and Assert to our work. We have applied test doubles using the Mockito library to stand in for external systems, simplifying the development.

In this book, we have covered a wide range of TDD and software design techniques. We can now create code with fewer defects, and that is safer and easier to work with.

Questions and answers

1. What further work could be done?

    Further work could include adding a **Continuous Integration (CI)** pipeline so that whenever we commit code, the application gets pulled from source control, built, and all tests run. We could consider deployment and automation of that. One example might be to package up the Wordz application and the Postgres database as a Docker image. It would be good to add database schema automation, using a tool such as Flyway.

2. Could we replace the Molecule library and use something else for our web endpoint?

    Yes. As the web endpoint sits in our adapter layer of the hexagonal architecture, it does not affect the core functionality in the domain model. Any suitable web framework could be used.

Further reading

- `https://martinfowler.com/articles/richardsonMaturityModel.html`

  An overview of what a REST web interface means, along with some common variations

- *Java OOP Done Right*, Alan Mellor, ISBN 9781527284449

  The author's book gives some more details on OO basics with some useful design patterns

- `https://www.postman.com/`

  A popular testing tool that sends HTTP requests and displays responses

- `http://molecule.vtence.com/`

  A lightweight HTTP framework for Java

- `https://undertow.io/`

  An HTTP server for Java that works well with the Molecule framework

- `https://github.com/google/gson`

  Google's library to convert between Java objects and the JSON format

- `https://aws.amazon.com/what-is/restful-api/`

  Amazon's guide to REST APIs

- `https://docs.oracle.com/en/java/javase/12/docs/api/java.net.http/java/net/http/HttpClient.html`

  Official Java documentation about the test HHTP client used in this chapter

Index

A

abstraction 19
acceptance testing tools 186, 187
adapters 156, 180
   replacing, with test doubles 166, 167
agile development 50
   combining, with TDD 52, 53
   exploring 50, 51
   user stories, reading 51, 52
American Standard Code for Information
      Interchange (ASCII) 95
Application Programming
      Interface (API) 49, 77
argument matchers 142-144
Arrange-Act-Assert (AAA) 56
automated code analysis
   benefits 204
   limitations 204
automated testing 202, 203, 221

B

bad code
   purpose 5
   writing 3, 4, 5
bad code, recognizing 6
   cohesion 10
   coupling 10
   error-prone constructs 9, 10
   high coupling, between classes 12
   low cohesion, in class 10, 11
   namespacing 7-9
   technical issues 6, 7
black box component 77
blue-green deployment 193
broken tests
   writing, concerns 35
build.gradle file
   libraries, adding 284
business outcomes
   diminishing 14, 15, 16

C

Characterization Test technique 39

Chicago TDD 223

**Chrysler Comprehensive
  Compensation project** 200

CI/CD pipelines 187, 190
  stages 190, 191

CI/CD workflows
  manual elements 209, 210

Classicist TDD 223

code
  documenting 28, 29
  test-driving, to play game 298-304

code-and-fix style 32

code coverage metric
  avoiding, as target 216, 217

code coverage tool 216

code readability 83

code review 203
  approaches 204, 205

code smell
  defining 82
  reference link 82

cohesion 10

collaborators
  error handling, testing challenges 121
  testing, challenges 120, 121
  unrepeatable behavior, testing
    challenges 120, 121

**Computer-Aided Software Engineering
  (CASE) tools** 215

Configuration Object 66

consumer-driven contract testing 183, 184

continuous delivery (CD) 190
  benefits 189
  defining 187
  need for 189

continuous deployment (CD) 190
  defining 187

continuous integration (CI)
  defining 187
  goal 188
  need for 187, 188

**continuous integration/continuous
  delivery (CI/CD) approach** 209

coupling 10

Cucumber
  URL 186

cyclomatic complexity (CYC) 21, 220

D

database
  designing 173
  word, fetching from 194-197

database adapters
  testing 181, 182

database integration test
  creating 268, 269
  production code 272-276

database port
  components 162

database-rider 195

DBRider
  database test, creating with 269-271

dead-code path 21

dependency injection 125

dependency inversion (DI)
  104, 125, 153, 154

Dependency Inversion Principle (DIP) 130
  applying, to shapes code 106-109
  irrelevant details, hiding 104-106

design

advancing, with two-letter
combinations 85-92

design flaws

revealing 22, 23

tests, writing benefits 23, 24

development environment

preparing 46

DevOps 208

diagnostic method 241

domain 20

domain layer

connecting to 290-293

domain model

calls, abstracting to web services 163, 164

code, writing 164

database, abstracting 162, 163

deciding 164, 165

frameworks, using 165

libraries, using 165

programming approach, deciding 165

web requests and responses,
abstracting 161, 162

E

end-to-end tests 177, 184-186

advantages and disadvantges 184

ensemble programming 205

error condition

testing, in Wordz 145-147

error handling code

with tests 144, 145

exceptions

asserting 63, 64

exploratory testing 202

external system

abstracting 160

domain model requirement 160

test doubles, substituting 166

external systems, challenges 150

accidental transactions 151

environmental problems 151

operating system calls 152

system time 152

third-party services 152, 153

uncertain data 152

Extreme Programming (XP) 28

F

failing test

writing 285-287

false negative test 180

**fast, isolated, repeatable, self-checking,
and timely (FIRST) 60, 61, 236**

feature flags 210

flaky tests 180

foreign method 241

functional programming (FP) 20, 158

future defects

protecting against 27, 28

fuzzing

reference link 208

G

getters 241

getXxx() methods 241

Goodhart's law 216

Google style guide

reference link 6

Google Tricorder 204

green phase 77, 78

H

happy path manual tests 5
Hawaii false missile attack warning
 reference link 152
hexagonal architecture 153-155
 adapters, testing 227
 domain logic, testing with 226, 227
 golden rule 159
 hexagon shape, need for 160
 test boundaries, defining with 226
 user stories, testing 228, 229
hexagonal architecture, components
 adapters, connecting to ports 157, 158
 external systems connect to adapters 156
 overviewing 155
 ports, connecting to domain model 158, 159
high cohesion 116
high coupling 12
HTTP server
 creating 288
 routes, adding 289, 290
Hypertext Markup Language (HTML) 99

I

information hiding 19
inside-out TDD approach 222
 advantages 223, 224
 challenges 224
 using 223
integration tests 177-180
 adapter layer 180, 181
 advantages 180
 consumer-driven contract testing 183, 184
 database adapters, testing 181, 182
 for database 194
 limitations 180
 web services, testing 182

IntelliJ IDE

IntelliJ IDE
 download link 46
 installing 46
Interface Segregation Principle (ISP)
 effective interfaces 115
 usage, reviewing in shapes code 115-117
Inversion of Control (IoC) 125
Invision
 URL 207

J

Java
 libraries and projects, setting up 46-48
Jest
 URL 206

K

keep it simple, stupid (KISS) 71

L

leaky abstractions
 preventing 25
legacy code 132
 managing, without tests 39
libraries
 adding, to project 284
Liskov Substitution Principle (LSP)
 swappable objects 109, 110
 usage, reviewing in shapes code 111
live-like environment 191
logic flaws
 manual testing, limits 26
 preventing 25
 test, automating to solve problems 26, 27
low cohesion 10

M

manual exploratory testing 201-203
manual testing
  limits 26
method name 19
Mockito 244
  argument matcher 142-144
  distinction, blurring between
    stubs and mocks 142
  URL 133
  used, for writing mock 141, 142
  used, for writing stub 133-141
  verify() method 142
  when() method 142
  working with 133
mock objects 127
  overuse, avoiding 130
  usage, scenarios 132
  using, to verify interactions 127-130
mocks
  code, avoiding for concrete class
    written outside of team 130, 131
  dependency injection 131, 132
  testing, avoidance 132
  value objects, avoiding 131

N

Net Promoter Score®™ (NPS) 14

O

object-oriented (OO) design 97
object-oriented programming
  (OOP) 20, 95, 158
object wiring 125

Open-Closed Principle (OCP)

adding, to new type of shape code 114, 115
extensible design 112, 113
open source HTTP libraries 284
outside-in TDD approach 224, 225
  advantages and disadvantages 225
OWASP Top 10 Web Application
  Security Risks
  reference link 208

P

Pact
  URL 184
pair programming 38, 205
penetration testing (pentesting) 208
Ports and Adapters 155
Postgres database
  installing 267, 268
pre-canned results
  stubs, using 126
procedural programming 20
public methods
  encapsulation, preserving 64, 65
  testing 64
pull model 127
pull request 204
push model 127

Q

Quality Assurance (QA) engineer 15
quality code
  abstraction 19, 20
  accidental complexity, avoiding 20-22
  designing 17, 18
  goal 18
  information hiding 19, 20
  naming 18, 19

R

RAID logs 215

random numbers adapters
  designing 173

red, green, refactor (RGR) cycle 75-79

refactor phase 78, 79

regression tests 201

repository 162

repository interfaces 168
  designing 169-173

RestAssured
  URL 186

RestEasy
  reference link 186

rider-junit5 library 269

S

sandbox APIs 183

scaffolding test 248

security risks 208

Selenium
  reference link 186

shapes code
  DIP, applying 106-109
  ISP usage, reviewing 115-117
  LSP usage, reviewing 111
  OCP, adding to type of 114, 115

shift-left approach 222

Single Responsibility Principle (SRP)
  blocks, building 97-104

SOLID principles 37, 97

Sonarqube
  URL 203

spike 40, 41

Spoofing, Tampering, Repudiation,
    Information Disclosure, Denial
    of Service, and Elevation of
    Privilege (STRIDE)
  reference link 208

Spring
  URL 125

staging environment 191

stub
  using, for pre-canned results 126
  writing, with Mockito 133-140

stub objects
  usage scenarios 127

stubs and mocks
  distinction, blurring between 142

switch-on-type 105

T

team performance
  decreasing 12, 13

technical debt 13

test boundaries
  defining, with hexagonal architecture 226

test doubles 122
  benefits 166
  production version of code,
      creating 124, 125
  purpose 122-124
  substituting, for external systems 166
  used, for replacing adapters 166, 167
  using, scenarios 130

test-driven application
  integrating 304-306

Test-Driven Development (TDD) 3, 31, 95
  agile development, combining with 52, 53
  automated regression testing 201
  benefits, of slowing down 32, 33

emergent design 217
expectations, managing 37
guide 96, 97
limits 200
myths 31-39
objections to not catching every
    bug, overcoming 35
objections to tests slowing users
    down, overcoming 33
problem-inflated expectations 36
test, writing before production code 40
test environments 181, 187, 191
advantages and disadvantages 191
test-first
acting, as design tool 214
Act stage 214
adding 213
Arrange stage 214
Assert stage 215
code coverage metrics 215, 216
form executable specifications 215
used, for making design decisions 214
writing, in continuous delivery
    situations 218
testing timeline 222
test-later
benefits 219
limitation 219-221
test pyramid 176-178
tests 65
act step 66
arrange step 65
assert step 66
automating, to solve problems 26, 27
code coverage 67
common errors, catching 62, 63
defining 59
error handling code with 144, 145

FIRST principles, applying 60, 61
missing 24
running, in production 192
single assert, using per test 61
writing, for Wordz application 79-85
tests, writing
after code, versus tests writing
    before code 218
benefits, analyzing before
    production code 23, 24
leaky abstractions, preventing 25
traffic partitioning 193, 194
triangulation 73, 86
two-letter combinations
design, advancing with 85-92

U

UML diagram
for shapes code 98
Unified Modeling Language (UML) 97
unit testing
bigger units 167
user stories 167, 168
unit tests 177-179
advantages 178
Arrange-Act-Assert (AAA) 56
domain model 179
limitations 66, 178
outcomes, working backwards 58
scope, deciding 61
structure, defining 56-58
workflow, increasing 59
untestable code
causes 37, 38
user acceptance tests 177, 184-186

user experience
    evaluating 207
user interface (UI) 27, 33
    testing 205, 206
user stories 50, 51
    reading 51, 52
    sections 52

V

value object 64, 131
variable name 19
verify() method 142

W

waterfall development 50
web services
    testing 182
when() method 142
WordRepository adapter
    implementation 276, 277
    database, accessing 277-279
    GameRepository, implementing 279
Wordz 168
    database, designing 173
    error condition, testing 145-147
    random numbers adapters, designing 173
    repository interface, designing 168-173
    writing 67-73
Wordz application 48
    rules, describing 49
    tests, writing 79-85
    word, fetching from database 194-197

Wordz game
    code, test-driving 298-304
    coding 235
    domain layer, connecting to 290-293
    ending 254
    errors, handling 294- 296
    failing test, fixing 296-298
    failing test, writing 285-287
    HTTP server, creating 288
    HTTP server routes, adding 289, 290
    libraries, adding to project 284
    progress, tracking 238-254
    score interface, designing 250-252
    start game code, refactoring 293
    starting 284
    test-driving 236, 237
    word selection logic, triangulating 244-249
Wordz game, end-of-game detection
    correct guess, responding 255, 256
    design, reviewing 260-263
    maximum number of guesses,
        triangulating 256, 257
    response, triangulating to guess
        after game over 257-259
Wordz web service application
    using, with Postman 306-309
wrong tests
    writing 67

Y

you ain't gonna need it (YAGNI) 71, 250

www.packtpub.com

Subscribe to our online digital library for full access to over 7,000 books and videos, as well as industry leading tools to help you plan your personal development and advance your career. For more information, please visit our website.

Why subscribe?

- Spend less time learning and more time coding with practical eBooks and Videos from over 4,000 industry professionals

- Improve your learning with Skill Plans built especially for you

- Get a free eBook or video every month

- Fully searchable for easy access to vital information

- Copy and paste, print, and bookmark content

Did you know that Packt offers eBook versions of every book published, with PDF and ePub files available? You can upgrade to the eBook version at packtpub.com and as a print book customer, you are entitled to a discount on the eBook copy. Get in touch with us at customercare@packtpub.com for more details.

At www.packtpub.com, you can also read a collection of free technical articles, sign up for a range of free newsletters, and receive exclusive discounts and offers on Packt books and eBooks.

Other Books You May Enjoy

If you enjoyed this book, you may be interested in these other books by Packt:

Pragmatic Test-Driven Development in C# and .NET

Adam Tibi

ISBN: 978-1-80323-019-1

- Writing unit tests with xUnit and getting to grips with dependency injection
- Implementing test doubles and mocking with Nsubstitute
- Using the TDD style for unit testing in conjunction with DDD and best practices
- Mixing TDD with the ASP.NET API, Entity Framework, and databases
- Moving to the next level by exploring continuous integration with GitHub
- Getting introduced to advanced mocking scenarios
- Championing your team and company for introducing TDD and unit testing

Test-Driven Development with C++

Abdul Wahid Tanner

ISBN: 978-1-80324-200-2

- Understand how to develop software using TDD
- Keep the code for the system as error-free as possible
- Refactor and redesign code confidently
- Communicate the requirements and behaviors of the code with your team
- Understand the differences between unit tests and integration tests
- Use TDD to create a minimal viable testing framework

Packt is searching for authors like you

If you're interested in becoming an author for Packt, please visit `authors.packtpub.com` and apply today. We have worked with thousands of developers and tech professionals, just like you, to help them share their insight with the global tech community. You can make a general application, apply for a specific hot topic that we are recruiting an author for, or submit your own idea.

Share Your Thoughts

Now you've finished *Test-Driven Development with Java*, we'd love to hear your thoughts! Scan the QR code below to go straight to the Amazon review page for this book and share your feedback or leave a review on the site that you purchased it from.

`https://packt.link/r/1-803-23623-X`

Your review is important to us and the tech community and will help us make sure we're delivering excellent quality content.